The **Intermediate** Guide to
Microsoft Excel 2007
Exam 77-602 Study Guide

another
Computer
Mama
Guide

	A	B	C	D	E
1	Initial	100			
2	Increment	5			
3					
4	Date	Product	Net	Quantity	Revenue
5	June 1, 2008	Pigs	$ 4.75	100	$ 475.00
6	June 2, 2008	Pigs	$ 4.75	=D5+$B$2	$ 498.75
7	June 3, 2008	Pigs	$ 4.75	105	$ 498.75
8	June 4, 2008	Pigs	$ 4.75	#VALUE!	#VALUE!
9	June 5, 2008	Pigs	$ 4.75	#VALUE!	#VALUE!
10	June 6, 2008	Pigs	$ 4.75	#VALUE!	#VALUE!

ISTE
This curriculum exceeds the National Education Technology Standards for Secondary Education

Microsoft
CERTIFIED
Application Specialist

Approved Courseware

© 2008 Comma Productions

The Comma Project

Intermediate Guide to Microsoft Excel 2007

© 2007 Comma Productions
9090 Chilson Road
Brighton, MI 48116
ISBN: 978-0-9818778-5-3

another
Computer Mama Guide

Trademark and Copyright
Microsoft Word ®, Microsoft Excel ®, Microsoft Access ®, Microsoft Outlook ®, Microsoft PowerPoint ®, Microsoft Windows ® are trademarks or registered trademark of Microsoft Corporation. Adobe Photoshop® is a trademark or register trademark of Adobe Corporation.

Limit of Liability/Disclaimer of Warranty:
The Publisher and Author make no representation or warranties with respect to the accuracy or completeness of the contents of this work and specifically disclaim all warranties including without limitations warranties of fitness for a particular purpose. The advice and strategies contained herein may not be suitable for every situation. The fact that an organization is refereed to in this work as a citation and/or potential source of further information does not mean that the authors or publisher endorses the information that the organization or website may provide or recommendations it may make. Readers should be aware that Internet websites listed in this work may have changed or moved between when this work was written and when it is read.

Neither Comma Productions nor Author shall be liable for any loss of profit or any other commercial damages including but not limited to special, incidental, consequential or other damages.

Comma Products

Comma Project, LLC.

Books Available in this Series:
Beginning Guide to Microsoft® Excel 2007

Intermediate Guide to Microsoft® Excel 2007

Advanced Guide to Microsoft® Excel 2007

About this Course
The Intermediate Guide to Microsoft Excel 2007
Exam 77-602: Using Microsoft® Office Excel 2007

Description: *The Intermediate Guide to Excel 2007* teaches how to use advanced text alignment, Conditional Formatting, create form controls, rename a worksheet, copy or move worksheets, link worksheets and consolidate data, use Relative and Absolute References, work with Scenarios, use Goal Seek, use the Auditing Toolbar.

Who will benefit from this course? Students who want to learn how to enter data into a spreadsheet and calculate the results will benefit from this course. Students who learn visually will appreciate the many illustrations and the step by step instructions. Successful students master how to create and edit formulas, audit equations, and format data.

Audience Description: Who should take this course?
The certification training is available to adult learners online in partnership with major colleges and universities. The audience for this course includes:
- Office workers, managers, entrepreneurs, teachers, and military personnel who want to start using advanced skills immediately
- Job training and professional development (WIA), as well as retired and unemployed people looking to expand their job possibilities

Course Prerequisites: Students who enroll in *The Intermediate Guide to Microsoft® Excel 2007* course should have basic computer skills including how to turn on the computer, how to use an Internet browser and how to select commands from a menu or toolbar. Students should also be proficient in file management: how to save, print and backup files.

© 2009 Comma Productions

Office Specialist Access Excel Outlook PowerPoint Word Vista

Online Course Requirements
Microsoft Certified Application Specialist (MCAS) certification training

You will need the following Microsoft products already installed on your computer in order to take this course online: Windows Vista Business edition, or Windows XP, For the Microsoft Office 2007 certification course, you MUST have the following software: Word 2007, Excel 2007, Access 2007; Outlook 2007and PowerPoint 2007. Microsoft Office 2007 is NOT the same as Microsoft Office 97-2003.

Hardware requirements for Vista Business:
- IBM-compatible (PC) computer running
- Processor: 1 GHz 32-bit (x86) or 64-bit (x64)
- RAM: 1 GB of system memory **(needs more)**
- Hard Drive: 40 GB with at least 15 GB of available space
- Video: Support for DirectX 9 graphics with WDDM Driver
- 128 MB of graphics memory (needs more)
- Pixel Shader 2.0 in hardware
- 32 bits per pixel
- DVD-ROM drive
- Audio Output

Adobe Acrobat Reader (free version) and a Flash Player.

An online course requires reliable, effect Internet access. If your internet service provider uses only dial-up, a minimum of 56K connection rate is recommended; however, high speed access (Cable or DSL) is preferred. This course cannot be taken with a Macintosh computer

Office Specialist Access Excel Outlook PowerPoint Word Vista

 Microsoft Certified Application Specialist

What is the Microsoft Business Certification Program?

Application Specialist

The Microsoft Business Certification Program enables candidates to show that they have something exceptional to offer – proven expertise in Microsoft Office programs. The two certification tracks allow candidates to choose how they want to exhibit their skills, either through validating skills within a specific Microsoft product or taking their knowledge to the next level and combining Microsoft programs to show that they can apply multiple skill sets to complete more complex office tasks. Recognized by businesses and schools around the world, over 3 million certifications have been obtained in over 100 different countries. The Microsoft Business Certification Program is the only Microsoft-approved certification program of its kind.

Application Professional

What is the Microsoft Certified Application Specialist Certification?

For more information:
Microsoft Business Certification

The **Microsoft Certified Application Specialist** Certification exams focus on validating specific skill sets within each of the Microsoft® Office system programs. The candidate can choose which exam(s) they want to take according to which skills they want to validate. The available Application Specialist exams include:

Using Microsoft ®Windows Vista™
Using Microsoft® Office Word 2007
Using Microsoft® Office Excel® 2007
Using Microsoft® Office PowerPoint® 2007
Using Microsoft® Office Access 2007
Using Microsoft® Office Outlook® 2007

Please Note: Comma Project, LLC. is independent from Microsoft Corporation, and not affiliated with Microsoft in any manner. While the Complete Computer Guides may be used in assisting individuals to prepare for a Microsoft Business Certification exam, Microsoft, its designated program administrator, and Comma Project, LLC. do not warrant that use of this Complete Computer Guides will ensure passing a Microsoft Business Certification exam.

Office Specialist Access Excel Outlook PowerPoint Word Vista

Exam 77-602: Using Microsoft® Office Excel 2007

Microsoft Certified Application Specialist (MCAS) reference topics

Description: The Microsoft Certified Application Specialist program is the only comprehensive, performance-based certification program approved by Microsoft to validate business computer skills using Microsoft Windows Vista® and Microsoft Office® 2007 productivity software: Excel, Word, Power Point, Access, and Outlook.

The Beginning Guide to Microsoft Excel 2007 demonstrates: Format font styles (size, color, and styles), Apply number formats (currency, percent), Format borders and shading, Use AutoSum, Insert and delete rows or columns, Hide and unhide rows and columns, Set up headers and footers, Use Functions: Average and IF, Enter text, dates and numbers, Use AutoFill, Use the Chart Wizard.

The Intermediate Guide to Microsoft Excel 2007 teaches: Conditional Formatting, Merge and Center Cells, Create form controls, Rename a worksheet, Copy or Move worksheets, Link worksheets and consolidate data, Use References, Work with Scenarios, Use Goal Seek, Use the Auditing Toolbar to trace errors.

The Advanced Guide to Microsoft Excel 2007 demonstrates how to name and manage ranges, copy or move worksheets, link worksheets and consolidate data, work with Scenarios, use Goal Seek, use the Auditing Toolbar to trace errors, create a grand total from several spreadsheets, understand Absolute and Relative References, analyze data with a PivotTable, filter the data, create Subtotals, Group and Outline data, and use the Logical, Lookup, Math and Text Functions.

Microsoft Certified Application Specialist (MCAS) objectives for Excel 2007

Study Guides
Beginning Excel
Intermediate Excel
Advanced Excel

MCAS Excel Excel Beginning Excel Intermediate Excel Advanced

 Microsoft Excel 2007 Study Guide
Microsoft Certified Application Specialist (MCAS): Microsoft Excel 2007 Exam 77-602 Guide

1. Creating and Manipulating Data
Fill a series without formatting, 29
Copy a series, 73
Ensure Data integrity, 49
Restrict data using data validation, 35
Restrict the type of data entered, 38
Restrict to: Less than x, 50
Restrict to: Specified length, 50
Restrict the values entered, 39
Create drop-down lists, 39
Remove duplicate rows 41
Change view to normal, page layout, and page break preview, 20
Copy worksheets, 76
Reposition worksheets, 76
Rename worksheets, 33
Insert and delete worksheets, 79

2. Formatting Data and Content
Insert cells, rows, columns, 79
Insert above, below, left, or right, 31
Insert multiple rows or columns, 83
Format rows and columns, 19
Format all cells in a row or column, 72
Autofit row height and column width, 78
Apply number formats, 72
Format the date, 69
Create custom cell formats, 70
Format text: font, alignment, 48
Merge and split cells, 48
Add and remove cell borders, 32
Format data as a table, 8
Apply Quick Styles to tables, 15
Apply and change Quick Styles, 15
Add and remove header rows, 12
Band the rows or columns in a table, 16
Change banded rows to columns, 16
Add rows to a tables, 17
Change the total row function, 18
Insert and delete rows and columns, 19

3. Creating and Modifying Formulas
Using absolute references, 86
Mixed references, 86
Troubleshoot a formula, 79
Use data from other worksheets, 80
SUM, 30

4. Presenting Data Visually
Conditional formatting, 55
Edit a conditional formatting rule, 60
More than one rule, 56
Highlight, 44, Top and bottom rules, 56
Data bars, 57, Color scales, 58
Icon sets, 59
Insert pictures, 46, Modify pictures, 47
Sort and filter data, 6

5. Collaborating and Securing Data
Use Compatibility Checker, 63
Save to the Excel 97-2003 format, 63
Set print options, 23
Define the print area, 23
Move a page break, 22
Preview and change a page break, 22
Change the orientation 21
Scale worksheet to fit 24

Microsoft Office Beginning Excel Intermediate Excel Advanced Excel

Table of Contents
Intermediate Guide to Microsoft® Excel 2007

1. Guided Discovery
Read the lessons and try the demonstrations.

The Table is Set
Tables and Print Options Page 11
Introduction page 12
Create a list page 14
AutoFit the Columns page 15
Single and multilevel sorts page 16
Create a Table page 18
Use the Header Row page 20
Convert a Table to Text page 23
Quick Style Formats page 25
Add a Total Row page 27
Add to the Table page 29
Print Preview page 30
Page Break Preview page 32
Set Print Area page 33
Print to Fit page 34

Working Overtime
Create a time sheet Page 37
Set Up a Timesheet page 38
Calculate the Total page 40
Create a Reference List page 46
Make a Drop-Down List page 49
Use Conditional Formatting page 51
Data Integrity and Validation page 59
Conditional Formatting page 65
Rules Manager page 70
Find Special Cells page 72
Compatibility Checker page 73

Legs, Eggs, and Pigs in a Basket
Calculate Revenue Page 75
Overview page 76
Enter the Data page 77
Format the Date and Time page 78
AutoFill the Data page 81
Calculate Revenue page 84
Copy the Spreadsheet page 86
Change the Variables page 87
The Summary Spreadsheet page 89
Link Worksheets page 90
Use Relative References page 92
Troubleshoot the equation page 95
Use Absolute References page 96
Use the Auditing Toolbar page 97
What If Scenarios page 99
Use Goal Seek page 102

Microsoft Office Beginning Excel Intermediate Excel Advanced Excel

Table of Contents, continued
Intermediate Guide to Microsoft® Excel 2007

2. Downloads
Samples

3. Assessment
Performance
Multiple Choice

Resources
Lesson Plans
Exercises

Objectives
Exploring Strategies
The Excel Timesheet uses a reference cell to calculate how your time compares to a regular 40 hour week.

Did you work any overtime? If so, Conditional Formatting can make the cell big, bold and colorful...just in case your boss needed some help finding the right information.

Questions: Visualization helps focus on the relevant information. Does Conditional Formatting help you focus on specific information?

Self-Assessment

Skill Level-Beginning Excel	Mastered	Needs Work	Required for my job
Select rows, columns, and cells			
Edit text and numbers in a cell			
AutoFill a series			
Create a graph with the Chart Wizard			
Add numbers with the AutoSum tool			
Calculate a simple equation "by hand"			
Sort data in a list			
Print a spreadsheet			

Beginning Excel is recommended if you selected "needs work" on three or more skills.

Skill Level-Intermediate Excel	Mastered	Needs Work	Required for my job
AutoFill data and formulas			
Link Spreadsheets			
Use Relative and Absolute cell references			
Use conditional formatting			
Create a drop down list			
Use the auditing toolbar			

Intermediate Excel is recommended if you selected "needs work" on three or more skills.

Skill Level-Advanced	Mastered	Needs Work	Required for my job
Create a Pivot Table for data analysis			
Format, group and graph Pivot Tables			
Use the What If tool			
Use Goal Seeking			
Import from Access or another database			
Use HLOOKUP and VLOOKUP			

Advanced Excel is recommended if you selected "needs work" on three or more skills.

Page **1** 2 3 4 5 6 7 8 9 10 11 12 13 14 15 16 17 18 19 20 21 22 23 24 25

Won't You Be My Neighbor?
The Table is Set

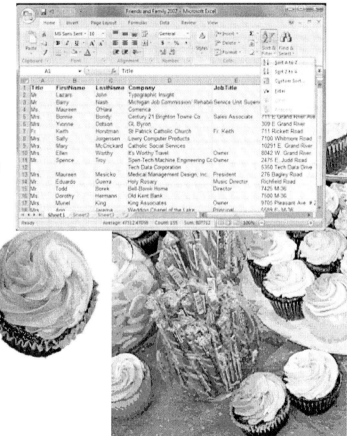

Click Here to Get Started
Sample Files

Intermediate Excel
Lesson Objectives: Learn how to create lists in Excel. This lesson shows how to present data visually, format data as a Table and set the print options. In this lesson you will:

Review how to create, format and sort a list page 4

Learn how to use single and multilevel sorts page 6

Explain the steps needed to create a Table page 8

Identify the Header Row with the Table Tools page 10

Locate the Quick Style Formats page 15

Use the Table Tools to add a Total Row page 17

Review the Print Preview options page 20

Locate and use Page Break Preview page 22

Explain the steps needed to Set Print Area page 23

Identify options to scale the spreadsheet to fit page 24

Excel: The Table is Set Page 1 2 3 4 5 6 7 8 9 10 11 12 13 14 15 16 17 18 19 20 21 22 23 24 25

Invite the Guests and Set the Table

Word and Excel are partners in efficiency. They work well together. Microsoft Word offers elegant **design** and **layout**. Microsoft Excel **organizes** and **sorts** the data.

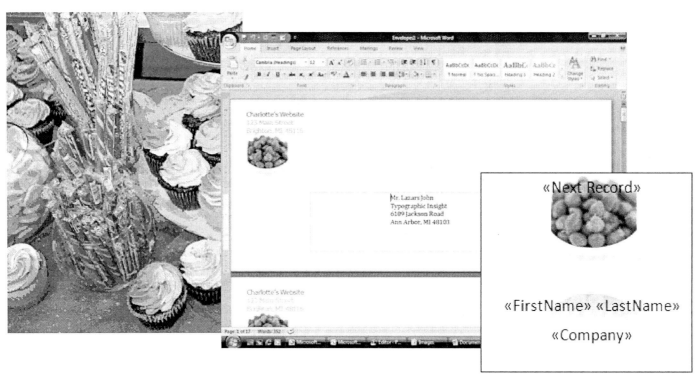

Excel: The Table is Set Page 1 2 3 4 5 6 7 8 9 10 11 12 13 14 15 16 17 18 19 20 21 22 23 24 25

Our goal is to create envelopes and labels for the Open House Flyer we designed in the first exercise. This work illustrates a common business process. In many, many companies, the clients names and addresses are recorded in a financial database. That information can be exported as a spreadsheet and used in a Mail Merge for a marketing campaign.

Start -> All Programs ->Microsoft Office-> Microsoft Office Excel 2007

Did Microsoft Excel Open? Yes.

Can you see the **Home** Ribbon? Yes? Good.

Excel: The Table is Set Page 1 2 3 4 5 6 7 8 9 10 11 12 13 14 15 16 17 18 19 20 21 22 23 24 25

Creating Lists in Excel

How do you make a record of all your clients for marketing and sales? Let's begin with the basics. There are three parts of any spreadsheet: labels, data, and formulas.

Try it: Enter the Labels
Please type: Last Name, (then tab to the next cell across) Title, (tab) First Name, (tab) Last Name, (tab)Company, (tab) JobTitle, (tab) Address1, (tab) Address2, (tab) City, (tab) State, (tab) Zip, (tab) Phone.

Did you notice there are two fields for the name? The more granular you make the data—the more detail you have to work with. For example, you could find customers with the same name but different addresses.

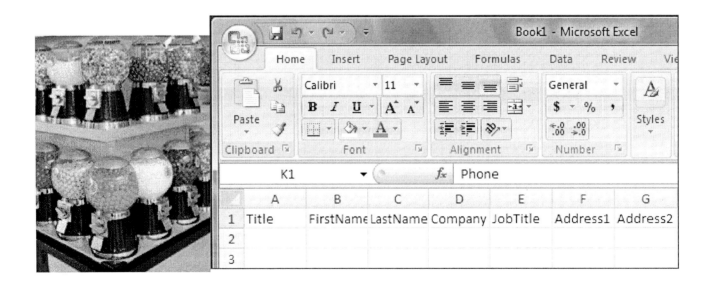

Excel: The Table is Set Page 1 2 3 4 5 6 7 8 9 10 11 12 13 14 15 16 17 18 19 20 21 22 23 24 25

Home -> Format -> AutoFit Column Width

AutoFit the Columns

Select the first row and all of the labels by clicking on the blue number one. This is the row selector. **Format** the labels to be bold by clicking the "**B**" on the **Home Ribbon**.

The bold formatting probably made the labels too wide for the columns. We can fix that:
1. **Select** the first row and all of the Labels
2. Go to **Format** on the **Home Ribbon**.
3. Select **AutoFit Column Width**.
Microsoft Excel makes the columns as wide as they need to be.

Memo to self: The information in Microsoft Outlook, Act! or Lotus Notes is real data. Each program has a method for exporting the files into Microsoft Excel. You can use the **sample spreadsheet** or create your own data, if you would like.

Microsoft Excel 2007 Exam 77-602 Topic: 2. Formatting Data and Content
2.2. Insert and modify rows and columns: Using AutoFit

Excel: The Table is Set Page 1 2 3 4 5 6 7 8 9 10 11 12 13 14 15 16 17 18 19 20 21 22 23 24 25

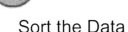

Home -> Sort & Filter -> Sort A to Z

Sort the Data

So here is a list of clients in Excel. What are the steps to sort the list by City and LastName?

Try it:
1. **Select** the spreadsheet. Click on the blue square at the top left hand corner where the rows and columns meet.

2. **Sort** the data. Go to the **Home** Ribbon and select **Sort & Filter**. Play with the options to Sort A-Z.

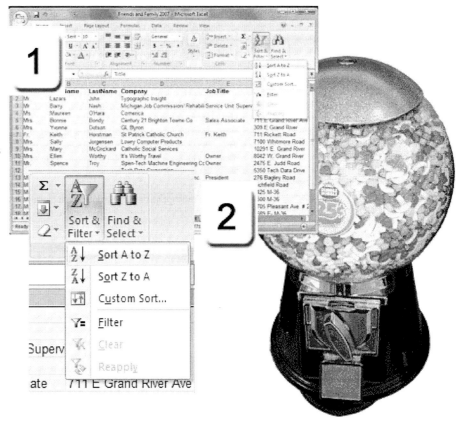

Microsoft Excel 2007 Exam 77-602 Topic: 4. Presenting Data Visually
4.6. Sort and filter data 4.6.1. Sort data using single or multiple criteria

Excel: The Table is Set Page 1 2 3 4 5 6 7 8 9 10 11 12 13 14 15 16 17 18 19 20 21 22 23 24 25

Home -> Sort & Filter -> Custom Sort

Sort the Data
This list has labels in the first row and that's called a **Header Row**. We can use the labels for sorting.

Try it:
Go to the **Home Ribbon**
Go to **Sort & Filter**
Select **Custom Sort**
Sort by the City
Click on **Add Level** to sort by LastName.

Give it a second and the entire list has been sorted. Not bad, huh?

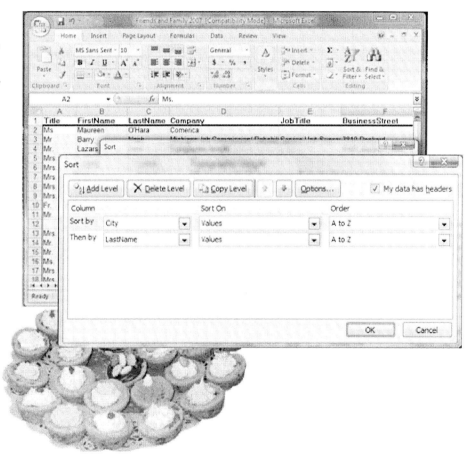

In the 1980s, Fr. Paul wanted to sort 550 families on a PDP 11 computer. It took most of the night. Computers are faster now, aren't they?

Excel: The Table is Set   Page 1 2 3 4 5 6 7 8 9 10 11 12 13 14 15 16 17 18 19 20 21 22 23 24 25

Working with Tables

One of the significant benefits of working with a computer is the ability to organize data into a Table: rows and columns of information. Here are the steps to convert a List into a Table.

Try It: Create a Table
1. Select the Range: Click on Cell A1 and highlight the labels and the data through Cell J18.

2. Insert a Table
Go to **Insert** ->**Table**.

You will be prompted. Go to the next page, please...

Insert->Table

Microsoft Excel 2007 Exam 77-602 Topic: 2.Formatting Data and Content
2.4. Format data as a table

Excel: The Table is Set Page 1 2 3 4 5 6 7 8 9 10 11 12 13 14 15 16 17 18 19 20 21 22 23 24 25

Insert->Table

Create a Table

The Table Wizard will display the **Range**. In the beginning of this sample, you selected Cell A1 through J18.

Note: the Range includes all of the labels and all of the rows of data.

My table has a header row.
A header row displays the **field names** at the top of the spreadsheet: for example last name, city, state, zip.

Memo to Self: $A$1 means **Absolute.** There is a detailed example of what this means later in this Guide.

*You can use the **sample spreadsheet** or enter your own names if you wish.*

Microsoft Excel 2007 Exam 77-602 Topic: 2.Formatting Data and Content
2.4. Format data as a table

Excel: The Table is Set Page 1 2 3 4 5 6 7 8 9 10 11 12 13 14 15 16 17 18 19 20 21 22 23 24 25

Table Tools->Design

View the Header Row

The header row is used to create **Subtotals** and **PivotTables** in Microsoft Excel.

The header row is used to create **Mail Merge Fields** in Microsoft Word, too.

What Do You See?
The Header Row is Bold. The Header Row has little arrows: drop-down option lists.

The rows of customer information have been formatted to make the data easier to read.

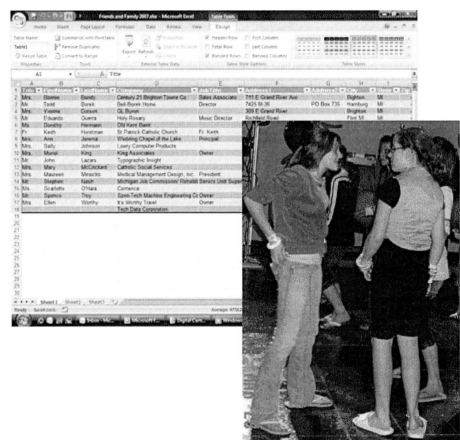

Excel: The Table is Set Page 1 2 3 4 5 6 7 8 9 10 11 12 13 14 15 16 17 18 19 20 21 22 23 24 25

Table Tools->Design

Use the Header Row

You can use the Header Row to **Sort** and **Filter** the data.

Try It: Apply Text Filters
Select the State Filter.
Sort the column A to Z.
Text Filters: Ann Arbor and Brighton.

What Do You See? This table is formatted with **Banded Rows**. Does the formatting update when you choose different Text Filters?

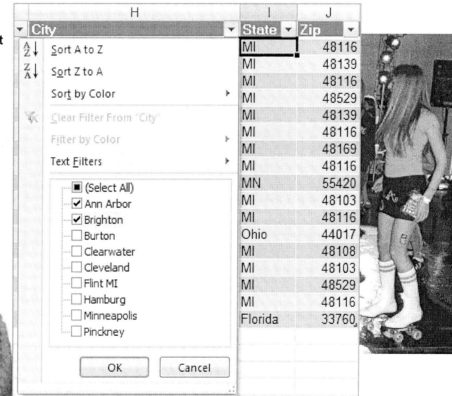

Microsoft Excel 2007 Exam 77-602 Topic: 2. Formatting Data and Content
2.4. Format data as a table: Add and remove header rows

Excel: The Table is Set Page 1 2 3 4 5 6 7 8 9 10 11 12 13 14 15 16 17 18 19 20 21 22 23 24 25

Table Tools->Design ->Table Style Options

Table Style Options
You can turn off the **Header Row** and hide all of the table functions associated with it, if you wish.

Try It: Add or Remove the Header
Go to **Table Tools-> Design**.
Look in the **Table Style Options**.
Uncheck the **Header Row**.

What Do You See?
When you turn off the Header Row, the first row becomes blank. The Header Row Field Names are saved in the table properties.

Trust But Verify: Turn on the Header Row, please.

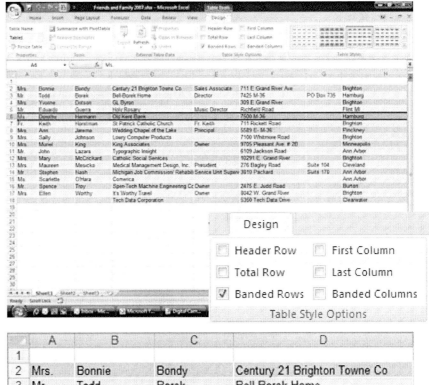

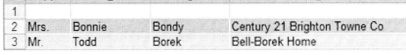

Microsoft Excel 2007 Exam 77-602 Topic: 2. Formatting Data and Content
2.4. Format data as a table: Add and remove header rows

Excel: The Table is Set Page 1 2 3 4 5 6 7 8 9 10 11 12 13 14 15 16 17 18 19 20 21 22 23 24 25

Table Tools->Design ->Tools

Convert Table to a Range

Each process in Microsoft has a **do** and an **undo**. How do you convert a table back to rows of data and remove the Header Row functionality?

Try It: Convert the Table to Text
Select the Table: A1:J18.
Go to **Table Tools -> Design**.
Select **Convert to Range**.

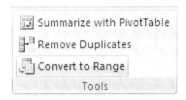

What Do You See? The Table is just rows of data, now, although the cells may still be formatted. The Header Row has lost the little filters.

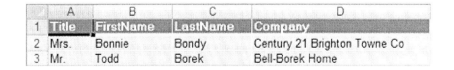

Microsoft Excel 2007 Exam 77-602 Topic: 2.Formatting Data and Content
2.4. Format data as a table

Excel: The Table is Set Page 1 2 3 4 5 6 7 8 9 10 11 12 13 14 15 16 17 18 19 20 21 22 23 24 25

Insert->Table

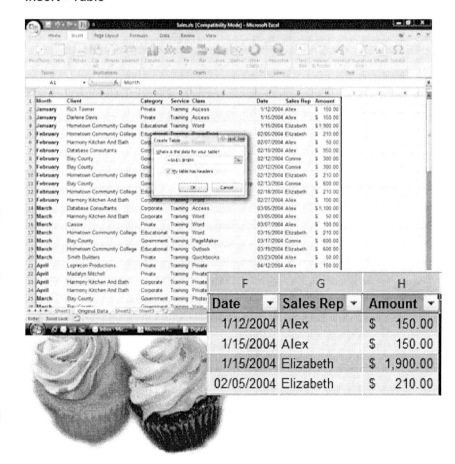

Different Data
Different Options

The next demonstrations show how to add a **Total Row** to a **Table** and how to change the **Summary** function.

Try It: Create a New Table
Open the sample file: Sales.xls.

You can use the sample spreadsheet or enter your own data if you wish.

Select the Range: A1 through H69
The key to working with a Total Row is to select just the data to create a new table, do not select the entire column or row.

Go to **Insert -> Table**.
Confirm the Range: =$A$1:$H$69
My list has Header Rows: Check

You should see a table that includes drop down combo boxes in the Header Row.

Microsoft Excel 2007 Exam 77-602 Topic: 2.Formatting Data and Content
2.4. Format data as a table

Excel: The Table is Set Page 1 2 3 4 5 6 7 8 9 10 11 12 13 14 **15** 16 17 18 19 20 21 22 23 24 25

Table Tools->Design ->Table Styles

Quick Style Formats

The purpose of formatting is to make the data easier to read and understand. The default table formatting in Microsoft Excel 2007 is very dark. There are better options. Microsoft Excel 2007 has a gallery of **Quick Style** templates. The Quick Styles format the Headers as well as the Columns and Rows.

Try It: Apply a Quick Style
Please click anywhere on the Table to select it. The **Table Tools Ribbon** should be visible.

Go to **Table Styles**.
Select a **Quick Style** from the gallery.

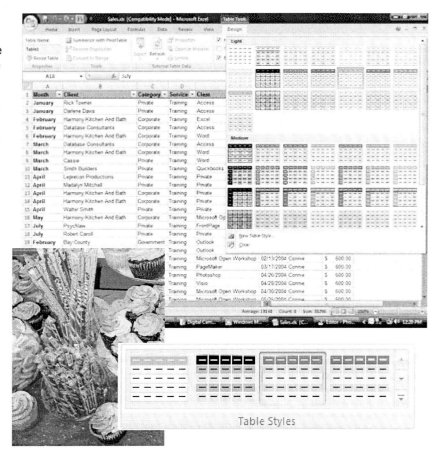

Microsoft Excel 2007 Exam 77-602 Topic: 2. Formatting Data and Content
2.4. Format data as a table: Apply and change Quick Styles

Excel: The Table is Set   Page 1 2 3 4 5 6 7 8 9 10 11 12 13 14 15 16 17 18 19 20 21 22 23 24 25

Table Tools->Design ->Table Style Options

Table Style Options

Color is also an effective method for emphasizing which data should get the most attention. In this sample, the data makes more sense when it is formatted as columns, not rows.

Try It: Edit the Table Style
Please click anywhere on the Table to select it. The **Table Tools Ribbon** should be available now.

Go to **Table Styles**.
Select: **Banded Columns**.
Select: First and Last Column.

What Do You See? The Table Styles emphasized the first and last columns by formatting the cells Bold.

The Banded Columns make it easier to see the data patterns in the Category, Services and Sales Rep Columns.

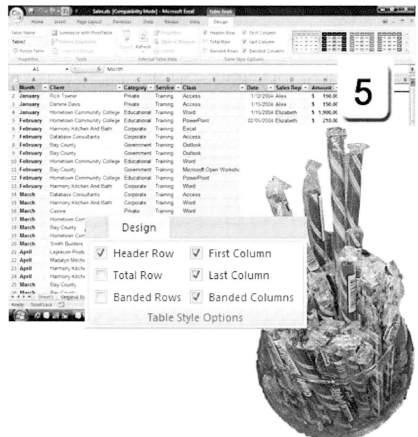

Microsoft Excel 2007 Exam 77-602 Topic: 2. Formatting Data and Content
2.4. Format data as a table: Band the rows or columns in a table (using banded rows)

Excel: The Table is Set   Page 1 2 3 4 5 6 7 8 9 10 11 12 13 14 15 16 17 18 19 20 21 22 23 24 25

Table Tools->Design ->Table Style Options -> Total Row

Calculate the Totals
Adding a Total Row to the Table is as simple as checking the option.

Try It: Add a Total Row
Go to **Table Styles**.
Select: **Total Row**.

What Do You See? By default, the Total Row will be added to the bottom of the Table. Microsoft Excel will insert a formula for your Total: $33,540.00.

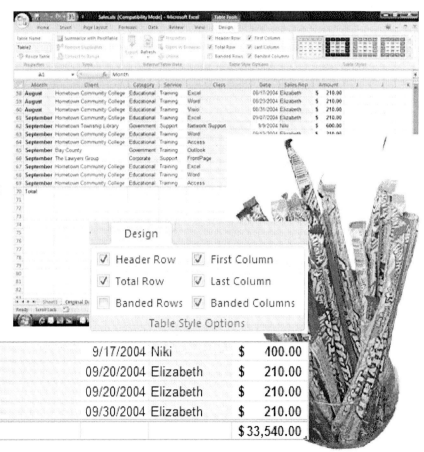

Microsoft Excel 2007 Exam 77-602 Topic: 2. Formatting Data and Content
2.4. Format data as a table   2.4.2. Add total rows to a tables

Excel: The Table is Set Page 1 2 3 4 5 6 7 8 9 10 11 12 13 14 15 16 17 18 19 20 21 22 23 24 25

Table Tools->Design ->Table Style Options -> Total Row

Edit the Total Row

Look at the Total Row: it has a summary function. You can change the operation from Sum to Average, Count, Max, Min or any of the options available in the extensive Excel library.

Try It: Edit the Total Row Function
Select: Cell H70. In this example, it is the cell with the Total formula.

Select a Function: Average.

What Do You See? The average price will be calculated. How would someone reading your spreadsheet know that this is the Average, and not the Sum? Labels, Labels, Labels.

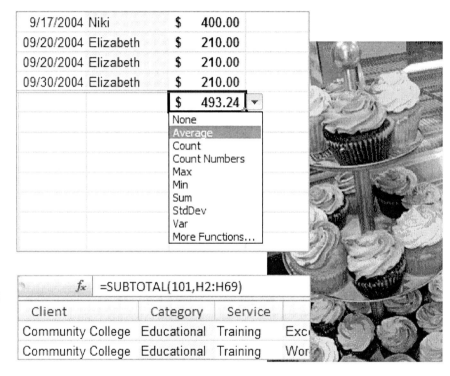

Microsoft Excel 2007 Exam 77-602 Topic: 2. Formatting Data and Content
2.4. Format data as a table 2.4.2. Change the summary function of total rows in tables

Excel: The Table is Set Page 1 2 3 4 5 6 7 8 9 10 11 12 13 14 15 16 17 18 19 20 21 22 23 24 25

Insert->Insert Table Rows Above

Add to the Table

Is the Table flexible? Can you add another row or column to the table without losing the formatting, totals calculations, or otherwise breaking it?

Try It: Add a Row
Select the Total Row.
Go to **Insert -> Table Rows Above**.

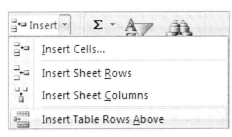

What Do You See? After you insert a new row or column, you may see a small format painter. If you click on the arrow, you can choose how you would like to format the new row.

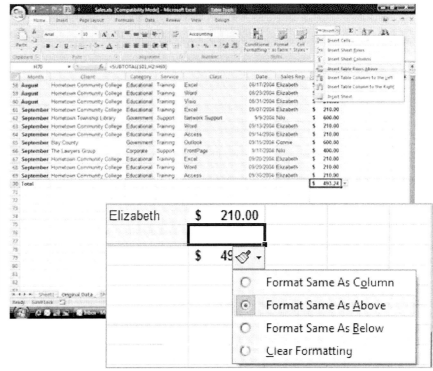

Excel: The Table is Set Page 1 2 3 4 5 6 7 8 9 10 11 12 13 14 15 16 17 18 19 20 21 22 23 24 25

Office->Print Preview

Print the Table
The Table is set. Take it into Print Preview and see if this spreadsheet is good enough for your company.

Try It: Print Preview
Go to **Office->Print Preview**.

What Do You See? Hmmmmm....

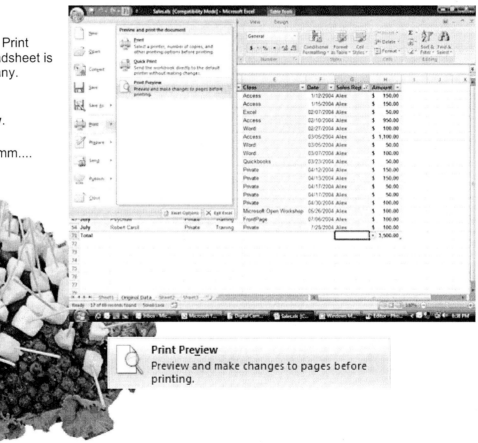

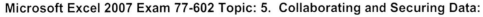

Microsoft Excel 2007 Exam 77-602 Topic: 5. Collaborating and Securing Data:
5.5. Set print options for printing data, worksheets and workbooks
5.5.2. Insert and move a page break: Preview and change a page break

Excel: The Table is Set Page 1 2 3 4 5 6 7 8 9 10 11 12 13 14 15 16 17 18 19 20 21 22 23 24 25

Page Layout> Orientation ->Landscape

Change the Orientation
Ok, Ok. Most spreadsheets are wide, not tall. In this Print Preview, you can see that all of the columns do not fit on one page when it is setup tall: Portrait.

Try It: Edit the Page Layout
Close the Print Preview.
Go to Page **Layout ->Orientation**.
Select **Landscape**.

What Do You See? After you look at a spreadsheet in Print Preview, you may see the Page Breaks as dashed lines along the columns and rows.

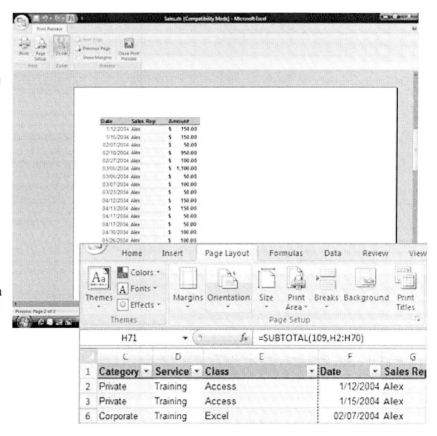

Microsoft Excel 2007 Exam 77-602 Topic: 5. Collaborating and Securing Data
5.5. Set print options for printing data, worksheets and workbooks
5.5.5. Change the orientation of a worksheet

Excel: The Table is Set Page 1 2 3 4 5 6 7 8 9 10 11 12 13 14 15 16 17 18 19 20 21 22 23 24 25

View -> Page Break Preview

Page Break Preview
There is another way to preview and adjust how the spreadsheet prints.

Try It: Edit the Page Breaks
Go to **View ->Page Break Preview**.

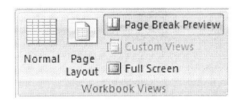

What Do You See? There will be a dashed line showing where the current page break is.

In this example, you could drag the dashed line between Column G and H to the right of Column H so that all of the columns will print on the same page.

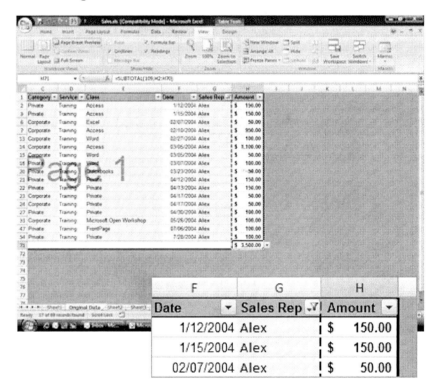

Microsoft Excel 2007 Exam 77-602 Topic: 5. Collaborating and Securing Data:
5.5. Set print options for printing data, worksheets and workbooks
5.5.2. Insert and move a page break: Preview and change a page break

Excel: The Table is Set Page 1 2 3 4 5 6 7 8 9 10 11 12 13 14 15 16 17 18 19 20 21 22 23 24 25

Page Layout->Print Area ->Set Print Area

Set the Print Area

Say there was only one piece in a big spreadsheet that you wanted to print. Here are the steps to do that.

Try It: Edit the Print Area
Select the Table: All of the Labels and all of the Data.

Go to **Page Layout -> Print Area**.
Click on **Set Print Area.**

What Do You See? The spreadsheet should have dashed lines around the range that you selected.

You can use the same steps to Clear a Print Area, too.

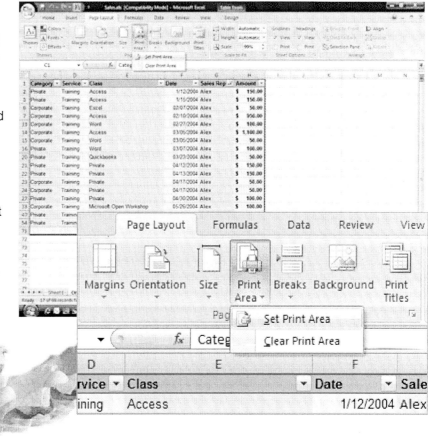

Microsoft Excel 2007 Exam 77-602 Topic: 5. Collaborating and Securing Data
5.5. Set print options for printing data, worksheets and workbooks
5.5.1. Define the area of a worksheet to be printed

Excel: The Table is Set Page 1 2 3 4 5 6 7 8 9 10 11 12 13 14 15 16 17 18 19 20 21 22 23 24 25

Page Layout->Page Setup

Print to Fit

There are a couple tricks that you can use to make a spreadsheet fit on a page.

Try It: Adjust the Print Scale
Go to **Page Layout ->Scale to Fit**.

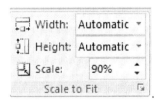

What Do You See? The toolbar has options for adjusting either the Width or the Height to fit one or more pages. You can also **Scale** the whole spreadsheet to print smaller.

Memo to Self: All of these settings can be found by clicking on the small arrow at the bottom right of the Scale to Fit group.

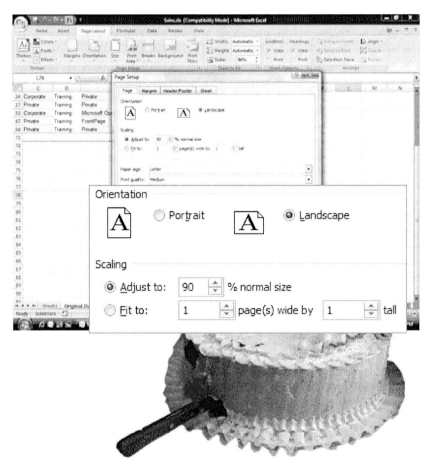

Microsoft Excel 2007 Exam 77-602 Topic: 5. Collaborating and Securing Data
5.5. Set print options for printing data, worksheets and workbooks
5.5.6. Scale worksheet content to fit a printed page

Excel: The Table is Set Page 1 2 3 4 5 6 7 8 9 10 11 12 13 14 15 16 17 18 19 20 21 22 23 24 25

DONE

From Soup to Nuts

This lesson illustrated how to create Tables. The Tables included a Header Row that allowed you to Sort and Filter the information.

In addition to using Quick Styles to format the Table, a Total Row was added and changed to show the Average.

Done and Done. We be good cuz we be smart! You can have two cookies. <grin>

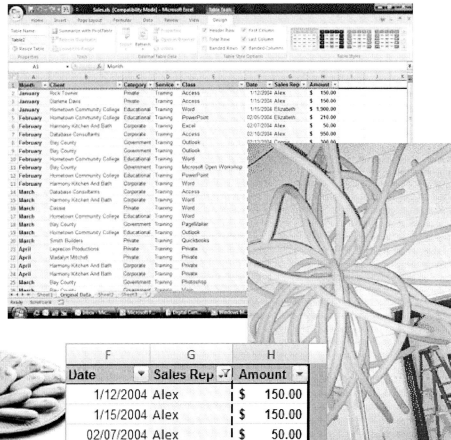

Test Yourself

1. Microsoft Excel can convert a list into a Table.
a. TRUE
b. FALSE
   Tip: Intermediate Excel, page 18

2. What are Banded Rows in Table?
a. The rows alternate colors to be easier to read
b. The rows alternate between bold and normal font to be easier to read
c. The rows are shaded in one color
   Tip: Intermediate Excel, page 22

3. The Header Rows in an Excel table include drop down menus for added functionality, including sorting.
a. TRUE
b. FALSE
   Tip: Intermediate Excel, page 22

4. Does Excel have a command to convert a Table to a Range?
a. Yes
b. No
   Tip: Intermediate Excel, page 23

5. Which of the following parts of a table are formatted using Quick Styles? (Select all correct answers)
a. Headers
b. Columns
c. Rows
d. Page Width
e. Font
   Tip: Intermediate Excel, page 25

6. Which command shows how to insert a Total Row in an Excel Table?
a. Table Tools-> Table Styles, click the Total Row checkbox
b. Insert-> Total Row
c. Format-> Table Styles-> Total Row
   Tip: Intermediate Excel, page 27

7. Rows or columns CANNOT be added to an Excel Table.
a. TRUE
b. FALSE
   Tip: Intermediate Excel, page 29

8. Set Print Area is used to select one area of a spreadsheet to print.
a. TRUE
b. FALSE
   Tip: Intermediate Excel, page 33

9. Which of the following can be adjusted with Scale to Fit commands? (Select all correct answers)
a. Width of the spreadsheet
b. Height of the spreadsheet
c. Scale of the spreadsheet
d. Orientation of the Spreadsheet
   Tip: Intermediate Excel, page 34

Page 1 2 3 4 5 6 7 8 9 10 11 12 13 14 15 16 17 18 19 More

  Does Anyone Really Know What Time It Is?
Working Overtime

Click Here to Get Started
Sample Files

Intermediate Excel

Lesson Objectives: Learn how to create a time sheet that uses client names in a drop down list. This lesson shows how to create and modify formulas, format content, ensure data integrity and restrict data using data validation. This lesson also demonstrates how to use Conditional Formatting to highlight specific information. In this lesson you will:

Practice how to enter data and fill a series page 2

Review the steps needed to calculate the Total page 4

Practice how to enter data in a reference list page 15

Identify the steps needed to make a drop-down list page 18

Learn how to use Conditional Formatting page 20

Identify options for Data Integrity and Validation page 21

Investigate Conditional Formatting Rules page 29

Use the Home Ribbon to Find Special Cells page 36

Learn how to use the Compatibility Checker page 37

Excel: Overtime Page 1 2 3 4 5 6 7 8 9 10 11 12 13 14 15 16 17 18 19 20 More

Create A Time Sheet In Excel

Are you a clock-watcher? I am. Time is money. The Computer Mama gets paid by the hour, so she has to keep track of how long it takes to help folks with their computers. The technical term is billable hours. Creating time sheets is a slow, repetitive job. Let me show you a way to do it faster with Excel. **Start** the **Program** Microsoft **Excel**.

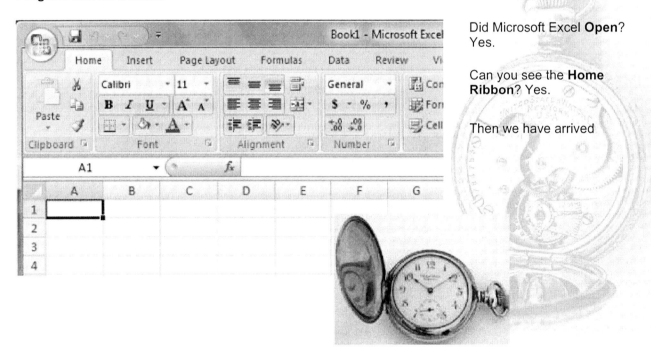

Did Microsoft Excel **Open**? Yes.

Can you see the **Home Ribbon**? Yes.

Then we have arrived

Excel: Overtime Page 1 2 **3** 4 5 6 7 8 9 10 11 12 13 14 15 16 17 18 19 20 More

Begin with the Labels

There are three parts to a spreadsheet: labels, data, and formulas. Let's begin with the labels.

Type "Monday" in cell A1. See the little **AutoFill** handle in the bottom right hand corner of the cell? Hover the cursor over that AutoFill handle until it changes into a small, black plus sign. Now, hold your left mouse button and drag to the right. Excel will fill in the series.

OK. We've got Monday through Saturday. Type "Total" in the cell next to Saturday. That's it for labels.

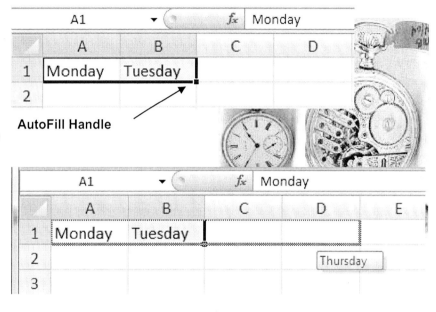

AutoFill Handle

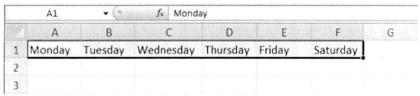

Monday, Monday! Can't Trust That Day.

Excel: Overtime Page 1 2 3 4 5 6 7 8 9 10 11 12 13 14 15 16 17 18 19 20 More

=Sum(A2:F2)

Calculate the Total

1. We need an equation in cell G2 to add up all of our hours. All good equations start with "=."

Type **=SUM(** into cell G2

2. Now select from A2 through F2. The formula says, =SUM(A2:F2.

Finish it off with the closed parenthesis. Finally, hit the **Enter** key. Excel won't calculate until you leave the cell.

Verify your work! Pretend we worked 8 hours on Monday, Wednesday and Friday. Does the Total equal 24?

AutoFill the Total
Click once on cell G2 to select the formula we made for the Total.

Use the AutoFill handle to copy the equation down to Row 12.

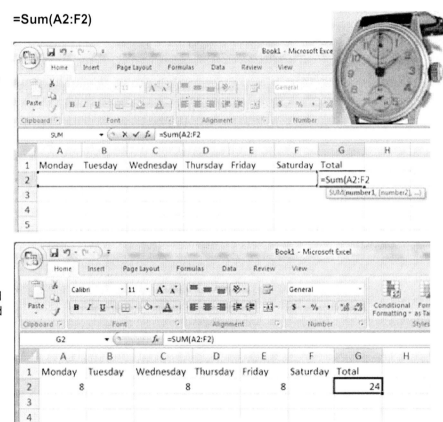

Microsoft Excel 2007 Exam 77-602 Topic: 3. Creating and Modifying Formulas
3.2. Summarize data using a formula 3.2.1 Use SUM, COUNT, COUNTA, AVERAGE, MIN, and MAX

Excel: Overtime Page 1 2 3 4 5 6 7 8 9 10 11 12 13 14 15 16 17 18 19 20 More

Home -> Cells -> Insert -> Insert Sheet Columns

Insert Columns

Suppose we work for more than one company. How can we expand our time sheet to show that?

Add a new column
Select column A by clicking on the "A" column header. Now, go to the **Home** Ribbon and look for the **Cells** group. Click on **Insert** on the menu bar, and then click on **Insert Sheet Columns**.

There will be a new, blank column A. Click on cell A1 and type: Accounts.

Labels should be Bold
Select Row 1 and make the labels bold by clicking the "B" on the formatting bar.

Labels should fit
Select Row 1.
Go to **Home ->Cells -> Format -> AutoFit Column Width**

Microsoft Excel 2007 Exam 77-602 Topic: 2. Formatting Data and Content
2.2. Insert and modify rows and columns: Insert a column or row above, below, to the left, or right

Excel: Overtime Page 1 2 3 4 5 6 7 8 9 10 11 12 13 14 15 16 17 18 19 20 More

Home ->Font ->Borders

Add Borders

Select from A1 to H13 and add some borders.

Remember, nothing happens on a computer until you select it, first.

The **Border** options are on the **Home Ribbon** in the **Font** group right next to the bucket of paint.

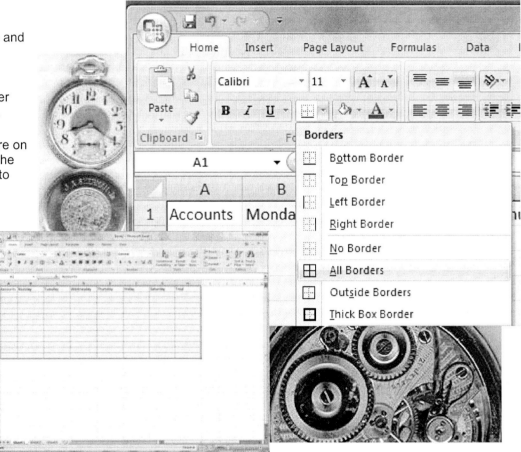

Microsoft Excel 2007 Exam 77-602 Topic: 2. Formatting Data and Content
2.3. Format cells and cell content 2.3.7. Add and remove cell borders

Excel: Overtime Page 1 2 3 4 5 6 7 8 9 10 11 12 13 14 15 16 17 18 19 More

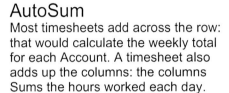

Home ->Editing ->AutoSum

AutoSum

Most timesheets add across the row: that would calculate the weekly total for each Account. A timesheet also adds up the columns: the columns Sums the hours worked each day.

Select B13. That is the cell where we want to add up the numbers.

Click on AutoSum button on the Home Ribbon Editing Group. Excel fills in the formula: =SUM(B2:B12)

Use Enter or Tab on your keyboard to exit the equation.

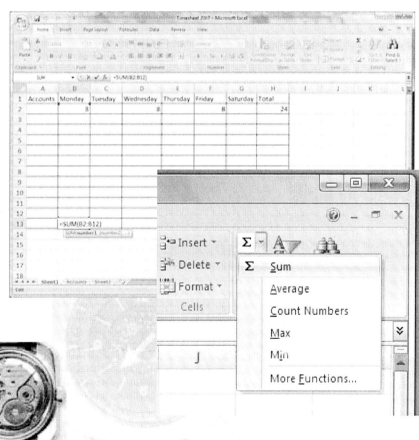

Microsoft Excel 2007 Exam 77-602 Topic: 3. Creating and Modifying Formulas
3.2. Summarize data using a formula 3.2.1 Use SUM, COUNT, COUNTA, AVERAGE, MIN, and MAX

Excel: Overtime Page 1 2 3 4 5 6 7 8 9 10 11 12 13 14 15 16 17 18 19 More

AutoFill the AutoSum
You can use the **AutoFill** handle to copy and paste the **AutoSum** formula into each column from Tuesday past Saturday to the Total column. Not bad: now we can see how many hours we worked for each day

A little more about AutoSum: Select the cell at the bottom of a column of numbers and click on AutoSum. Excel will fill in the equation for you:**=Sum(B2:B12)**

What happens if AutoSum does not select all of the cells you want? Click on the cell with the bad formula and look above the rows for the Formula Bar.

You can click in the formula bar and edit the equation to read: **=Sum(B2:B12)**

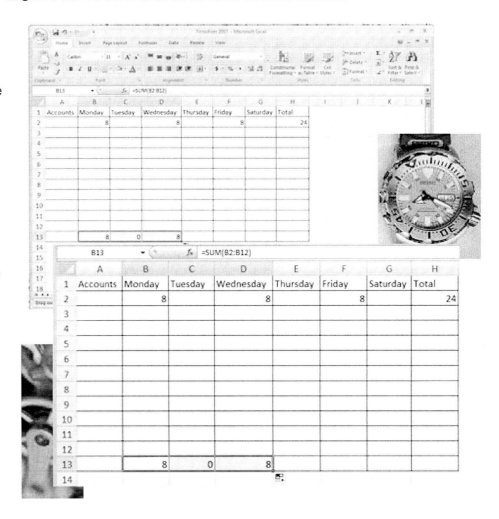

Microsoft Excel 2007 Exam 77-602 Topic: 3. Creating and Modifying Formulas
3.2. Summarize data using a formula 3.2.1 Use SUM, COUNT, COUNTA, AVERAGE, MIN, and MAX

Excel: Overtime Page 1 2 3 4 5 6 7 8 9 10 11 12 13 14 15 16 17 18 19 More

Calculating Overtime

Suppose we want this little time sheet to calculate the amount of overtime we worked.

Try it: Calculate the Total Hours
Select Cell H15 as a place to calculate the overtime.

The equation is =H13-H14, where H13 is the sum of all the hours worked minus H14, the cell that shows 40 hours in a regular week.

In our sample spreadsheet, we only worked 24 hours that week, and the equation shows we are 16 hours shy of a 40 hr week.

If our spreadsheet added up 45 hours this week, then the equation in cell H15 would show 5 hours of overtime.

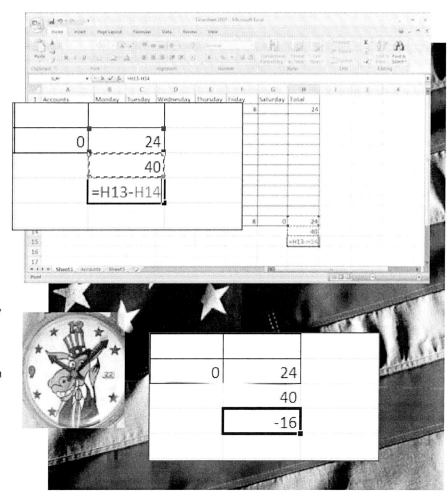

Excel: Overtime Page 1 2 3 4 5 6 7 8 9 10 11 12 13 14 15 16 17 18 19 More

Home -> Cells -> Insert -> Insert Sheet Row

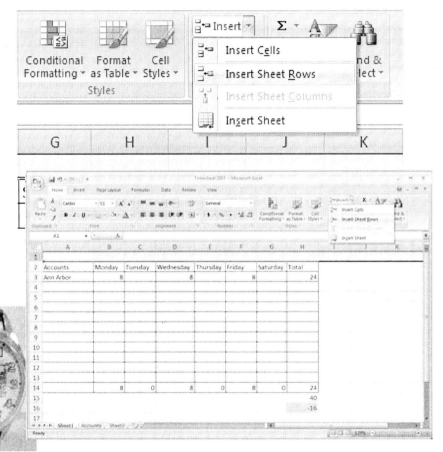

Adding Graphics
Microsoft Excel is just as rich in formatting and layout tools as Microsoft Word. However, most of the time it isn't appropriate to add graphics to the top row in a spreadsheet, especially if it is a list of data used in a Mail Merge.

This little time sheet is simple and adding some graphics would not interfere with the functions and formulas.

Start by adding a couple of extra rows at the top of the sheet. Select row 1 and go **Home** and look for the **Cells** Group. You should see the **Insert** button.

Microsoft Excel 2007 Exam 77-602 Topic: **2. Formatting Data and Content**
2.2. Insert and modify rows and columns: Insert a column or row above, below, to the left, or right

Excel: Overtime Page 1 2 3 4 5 6 7 8 9 10 **11** 12 13 14 15 16 17 18 19 More

Home -> Illustrations -> Picture

Working with Graphics

Using graphics in Microsoft Excel is similar to working with a table in Microsoft Word. We will learn more about Tables later.

You can add a computer logo by inserting a **Picture from ClipArt** or using a **Picture from File** and browsing for a sample picture.

Yes, you can use pictures on a simple spreadsheet. <grin>

Memo to self: Go easy on the graphics. Most spreadsheets will not function as a data source for a mail merge if there is anything above the header row in Row 1.

Microsoft Excel 2007 Exam 77-602 Topic: 4. Presenting Data Visually
4.4. Insert and modify illustrations 4.4.1. Insert and modify pictures from files (not clip art files)

Excel: Overtime Page 1 2 3 4 5 6 7 8 9 10 11 **12** 13 14 15 16 17 18 19 More

Home -> Illustrations -> Picture

More Formatting

You can **resize the picture** by dragging one of the handles until the picture is smaller.

What Do You See? Did you notice when you clicked on the picture, the Ribbon changed to the **Format** view and offered the **Picture Tools**?

You can **resize the rows** in a similar way. When you run your mouse over the line between the rows you will see a double-headed arrow. Hold your left mouse and drag the row taller.

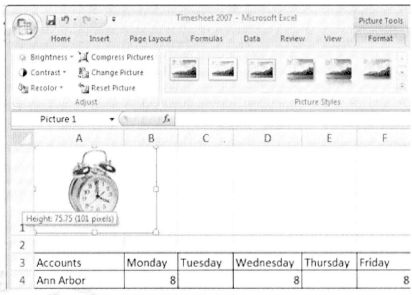

Microsoft Excel 2007 Exam 77-602 Topic: 4. Presenting Data Visually
4.4. Insert and modify illustrations 4.4.1. Insert and modify pictures from files (not clip art files)

Excel: Overtime Page 1 2 3 4 5 6 7 8 9 10 11 12 **13** 14 15 16 17 18 19 More

Home -> Alignment -> Merge and Center

Text Alignment
Look on the **Home** Ribbon for the **Alignment** Group. These are the tools you need to align the text left, right, or center. You can also format the text so it sits at the top, middle, or bottom of the cell.

Try this: Merge and Center
Select from C1 through H1.
Go to **Home -> Alignment -> Merge and Center.**

Enter the Label: Weekly Time Sheet

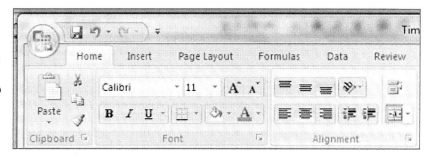
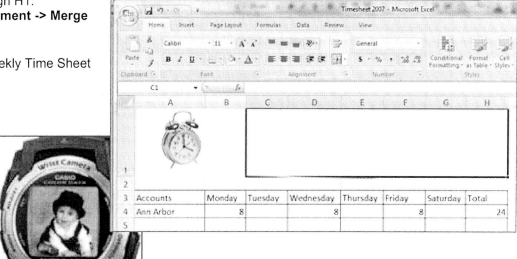

Microsoft Excel 2007 Exam 77-602 Topic: 2. Formatting Data and Content
2.3. Format cells and cell content   2.3.6. Merge and split cells

Excel: Overtime Page 1 2 3 4 5 6 7 8 9 10 11 12 13 **14** 15 16 17 18 19 More

Data -> Data Tools -> Data Validation

Adding Data in a Spreadsheet

This time sheet is a fast and effective way to record your hours. Suppose you work for the same Accounts every week. What if you could just choose the Account from a list? Choosing the answer from a list has two benefits: the data entry is faster, and it is spelled correctly.

Data Validation restricts what goes into a cell. For example, suppose a spreadsheet had a column for the **Date**.

If you use Data Validation, Excel would compare what the user types and alert the user if they typed in something wrong, like a price instead of a date.

Data Validation can also create a simple drop down List.

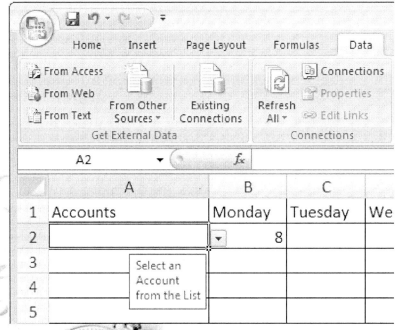

This screen shot is an example of the drop down combo box you will create in the next pages with Data Validation.

Excel: Overtime Page 1 2 3 4 5 6 7 8 9 10 11 12 13 14 <u>15</u> 16 17 <u>18</u> <u>19</u> More

Make a Reference List

There are two parts to making a drop down list. First you need a spreadsheet for the names on the list. Then you need to create a Drop Down List with Data Validation to display the names.

Start with the Accounts List

Select Sheet 2 in the workbook. Then add eight sample names to the list in cells A1 through A8.

Let's **rename** the tab at the bottom of the spreadsheet. Double click on the tab on the bottom. When the **label** for Sheet 2 is highlighted, type Accounts.

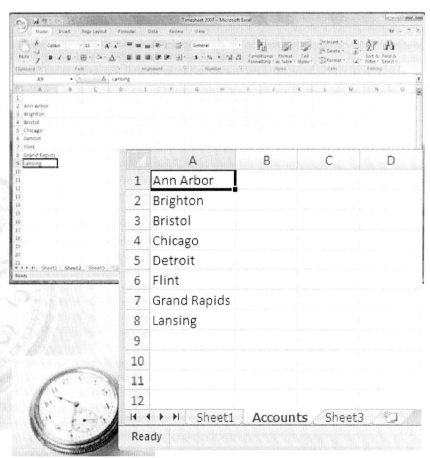

Excel: Overtime Page 1 2 3 4 5 6 7 8 9 10 11 12 13 14 15 16 17 18 19 More

Name the Data Source

The Drop Down List will use our Accounts spreadsheet as the data source. Here are the steps to prepare for the Data Validation.

1. Go to the Accounts spreadsheet and select cells A1 through A8.

2. Name that Range: Clients
The **Name Box** is located right above cell A1. Click in the Name box, type, then click Enter on your keyboard to save your settings.

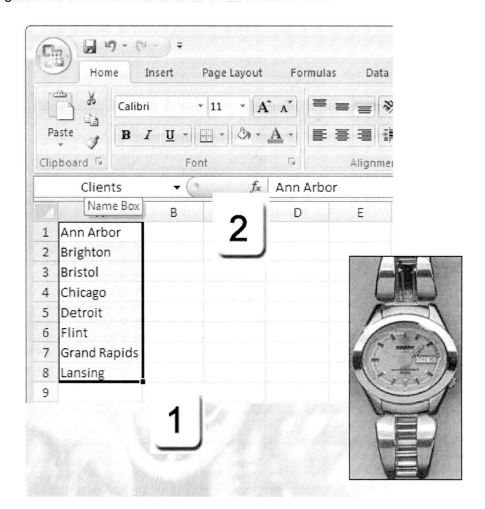

Microsoft Excel 2007 Exam 77-602 Topic: 1. Creating and Manipulating Data
1.2.1. Restrict data using data validation: Create drop-down lists

Excel: Overtime Page 1 2 3 4 5 6 7 8 9 10 11 12 13 14 15 16 17 18 19 More

Data -> Data Tools -> Data Validation

Validation Settings

Data Validation has several interesting format options. Go back to Sheet 1 and select cell A2 to begin.

Go to the **Data Ribbon** and look in the **Data Tools** for **Data Validation**.

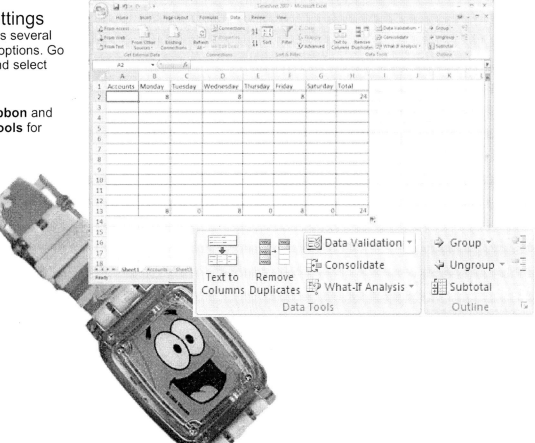

Excel: Overtime Page 1 2 3 4 5 6 7 8 9 10 11 12 13 14 15 16 17 18 19 More

Data -> Data Tools -> Data Validation

Validation Settings

The default value for a cell is Any Value. Validation criteria can restrict data entry to Whole numbers, decimals, Dates or Time.

For our example, select **List** from the options.

After you **Allow** the List, you need to identify the data **Source**. Our list is on the Accounts spreadsheet. We named that range: clients.

Click in the **Source** box and type: =clients

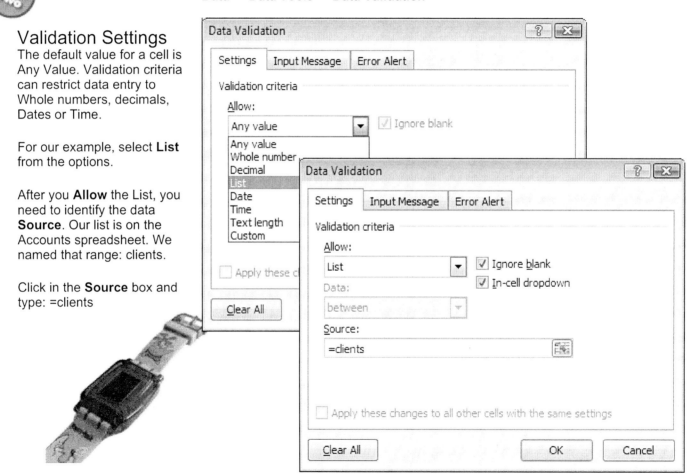

Microsoft Excel 2007 Exam 77-602 Topic: 1. Creating and Manipulating Data
1.2.1. Restrict data using data validation: Restrict the type of data that can be entered in cells

Excel: Overtime Page 1 2 3 4 5 6 7 8 9 10 11 12 13 14 15 16 17 18 **19** More

Input Message

This Validation option offers a method for teaching your users how to work with your form. Whatever you type here will pop up when a user clicks on the cell.

The Computer Mama has a theory that the more you help your users, instead of scaring the wits out of them, the more they will be inclined to use your forms.

Data ->Data Tools-> Data Validation

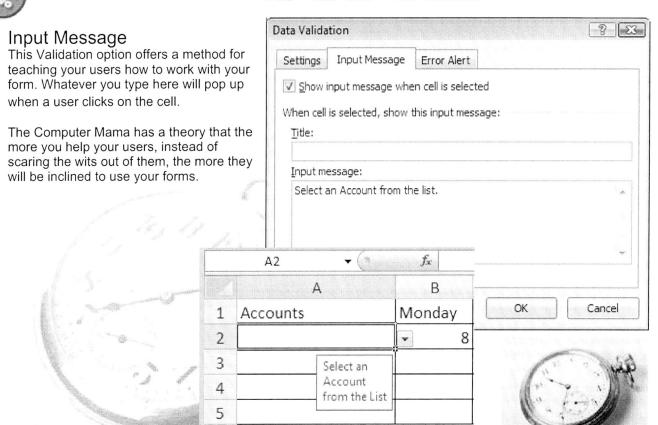

Microsoft Excel 2007 Exam 77-602 Topic: 1. Creating and Manipulating Data
1.2.1. Restrict data using data validation: Restrict the type of data that can be entered in cells

Excel: Overtime (Page 1) 19 20 21 22 23 24 25 26 27 28 29 30 31 32 33 34 35 36 37 38

Data ->Data Tools-> Data Validation

Error Message
An error message needs to be more than beeps and buzzers. Practical people want to know how to do the job right. Use the Error options to teach someone how to correct the mistake.

There are three **alert** styles : Stop, Warning and Information. For example, the Stop message has an annoying sound and too many buttons.

Try it: Edit the Validation Error
What is the difference between Retry and Cancel? Which choice deletes the wrong answer?

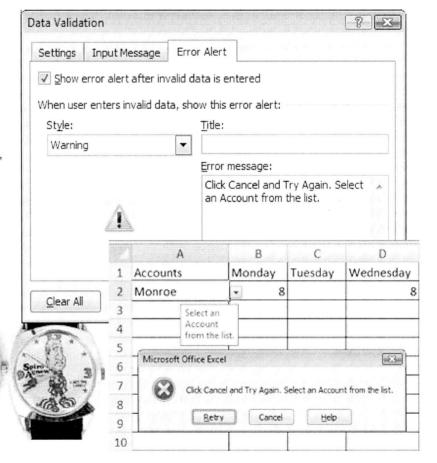

Microsoft Excel 2007 Exam 77-602 Topic: 1. Creating and Manipulating Data
1.2.1. Restrict data using data validation: Restrict the type of data that can be entered in cells

Excel: Overtime (Page 1) 19 20 21 22 23 24 25 26 27 28 29 30 31 32 33 34 35 36 37 38

Data -> Data Validation

Another Example Using Data Validation

You want your users to be successful when they fill in the weekly timesheet.

Data Validation can also be used to ensure data integrity. **Data Integrity** means the user typed in the value that belongs to the cell.

For example: did your users enter today's date instead of number of hours in the time sheet?

Instead of a combo box, this example will use an Input Message and Error Alerts to prompt users to enter the correct data.

Try It: Validate the Hours
Select: Cell B5.
Go to **Data -> Data Validation**.

Excel: Overtime (Page 1) 19 20 21 **22** 23 24 25 26 27 28 29 30 31 32 33 34 35 36 37 38

Data Validation

You can specify that the hours typed into the cells should be greater than 0 (zero) but less than 12. Here are the steps to restrict data entry.

Try It: Edit the Validation Criteria
Select: Cell B5.
Go to **Data -> Data Validation.**
Allow: Whole Number
Data: Between
Minimum: 0 (That's a zero.)
Maximum: 12

Keep going to the next page...

Data -> Data Validation -> Settings

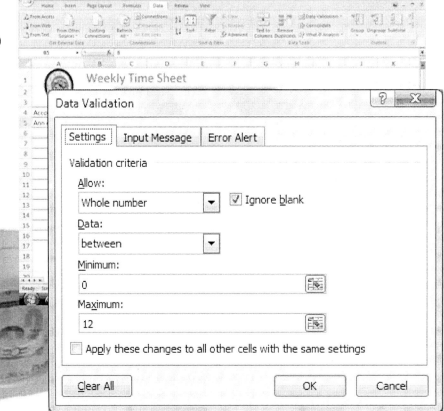

Microsoft Excel 2007 Exam 77-602 Topic: 1. Creating and Manipulating Data
1.2. Ensure data integrity: Less than x

Excel: Overtime (Page 1) 19 20 21 22 <u>23</u> 24 25 26 27 28 29 30 31 32 33 34 35 36 37 38

Data -> Data Validation -> Settings

Validation Settings

If you give a little, you get a lot. Use the **Input Message** and **Error Alert** to help your users use this Timesheet right.

Try It: Edit the Input and Error
Input Message: Enter the hours: 1-11.
Error Style: Information
Error Message: Click CANCEL and try again. You need to type a number greater than zero and less than 12 hours!

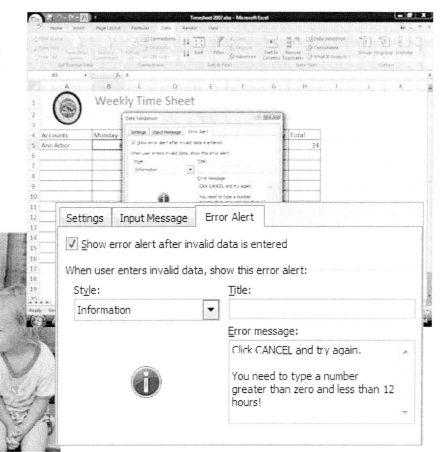

1.2. Ensure data integrity: Less than x

Excel: Overtime (Page 1) 19 20 21 22 23 24 25 26 27 28 29 30 31 32 33 34 35 36 37 38

Data -> Data Validation -> Settings

Test the Settings
You need to try the Data Validation and see if the settings work.

Try this: Test the Validation Settings
Select Cell B5.
Type: 8
Tab to the next cell.

What Do You See? Before you entered any data, did the cell display a yellow balloon with an Input Message? If you click on another cell in the spreadsheet, did the little Input Message hide?

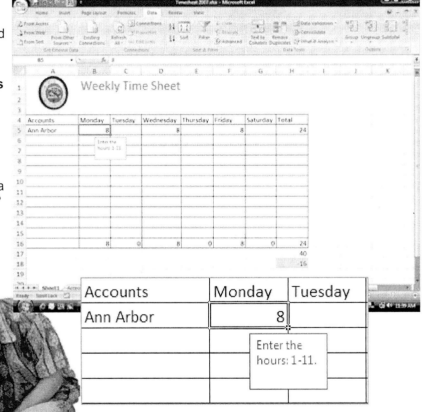

Microsoft Excel 2007 Exam 77-602 Topic: 1. Creating and Manipulating Data
1.2. Ensure data integrity: Less than x

Excel: Overtime (Page 1) 19 20 21 22 23 24 **25** 26 27 28 29 30 31 32 33 34 35 36 37 38

Data -> Data Validation -> Settings

Make a Mistake
Try this: Test the Error Alert
Select Cell B5.
Type: 13
Tab to the next cell.

What Do You See? The Error Alert asks for three different responses: OK, CANCEL or HELP. Which one should you choose? What is, "OK?"

If you have the patience to play with the different responses, you'll see that only CANCEL clears the cell and lets the user start over. So, that's why those instructions are included in your Error message.

What Do You Hear? The different alerts have different default sounds attached to them.

Excel: Overtime (Page 1) 19 20 21 22 23 24 25 26 27 28 29 30 31 32 33 34 35 36 37 38

Copy the Code

The Data Validation programmed in Cell B5 can be copied to the other cells in the spreadsheet.

Try this: Copy the Data Validation
Select Cell B5.
AutoFill: Cell C5:H15.

What Do You See? Click on any cell and test whether the **Data Validation** settings were copied?

If you enter a number greater than the Validation Rule, say 13, will you get an Error Alert?

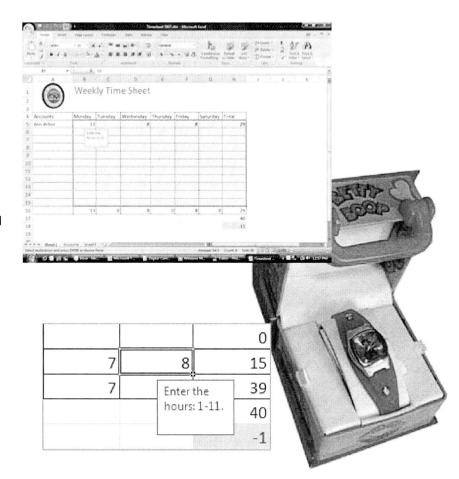

Excel: Overtime (Page 1) 19 20 21 22 23 24 25 26 <u>27</u> 28 29 30 31 32 33 34 35 36 37 38

Home ->Styles ->Conditional Formatting

Conditional Formatting

Conditional formatting is a way to change the color or size of a cell, based on the results of an equation.

Our timesheet is tracking overtime. What if the overtime hours showed up as big, bold and green?

Try it: Use Conditional Formatting
Select Cell H15
Then go to the **Home Tab**
Look for the **Styles Group**.
Click on **Conditional Formatting**.

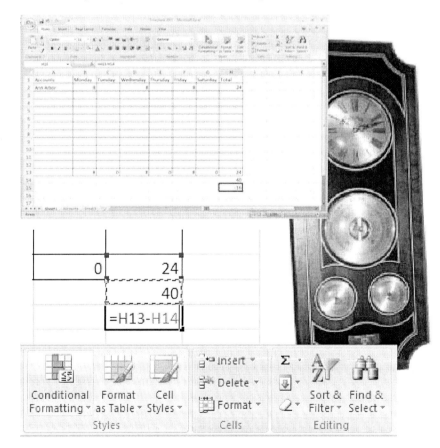

Microsoft Excel 2007 Exam 77-602 Topic: 4. Presenting Data Visually
4.3. Apply conditional formatting
4.3.1. Manage conditional formats by using the Conditional Formatting Rules Manager :Create a new rule

Excel: Overtime (Page 1) 19 20 21 22 23 24 25 26 27 28 29 30 31 32 33 34 35 36 37 38

Home ->Styles ->Conditional Formatting

Formatting Options

Here are the steps to create the Conditional Formats.

Cell Rules Greater Than 0:
Enter 0 (zero) for the amount and select Green Fill with Dark Green Text from the list and click on OK.

Cell Rules Less Than 0:
Repeat the experiment for results Less Than 0 (zero), but this time choose the red formatting option.

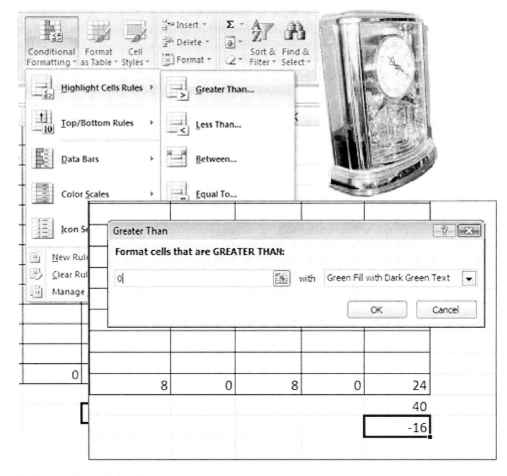

Microsoft Excel 2007 Exam 77-602 Topic: 4. Presenting Data Visually
4.3. Apply conditional formatting
4.3.1. Manage conditional formats by using the Conditional Formatting Rules Manager :Create a new rule

Excel: Overtime (Page 1) 19 20 21 22 23 24 25 26 27 28 <u>29</u> 30 31 32 33 34 35 36 37 38

Home -> Conditional Formatting -> Top/Bottom Rules

Conditional Formats

Conditional Formatting emphasizes data that meets certain criteria. It is a quick, efficient method for focusing attention.

The following illustrations use the sales spreadsheet. You can use the <u>sample spreadsheet</u> or enter your own data if you wish.

Try It: Above or Below Average?
Select Column H (In the sample spreadsheet, this is the column with the Amount).

Go to **Home ->Conditional Formatting**.
Go to **Top/Bottom Rules**.
Select: **Above Average.**

Format Cells: You will be prompted to select a formatting option. Select Green Fill with Dark Green Text, please.

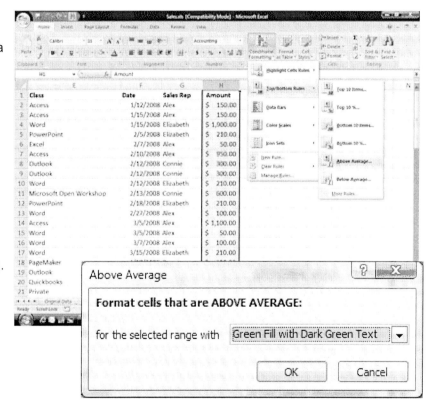

Microsoft Excel 2007 Exam 77-602 Topic: 4. Presenting Data Visually
4.3. Apply conditional formatting
4.3.1. Manage conditional formats by using the Conditional Formatting Rules Manager: Create a new conditional formatting rule

Excel: Overtime (Page 1) 19 20 21 22 23 24 25 26 27 28 29 30 31 32 33 34 35 36 37 38

Home -> Conditional Formatting -> Top/Bottom Rules

More Than One Rule
You can have more than one Conditional Rule formatting your data. In this illustration, there is one color for sales that are Above Average, and another for the sales that are Below Average.

Try It: Add Another Rule
Select Column H.
Go to **Home ->Conditional Formatting**.
Go to **Top/Bottom Rules**.
Select: **Above Average**.

Format Cells: You will be prompted to select a formatting option. Select Red Fill with Dark Red Text, please.

What Do You See? Any amount less than the Average is highlighted red with red text. Any amount greater than the Average is green/green.

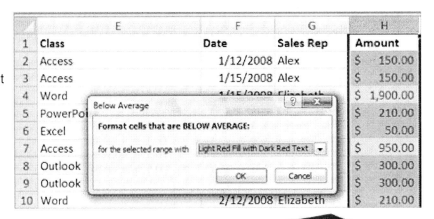

Microsoft Excel 2007 Exam 77-602 Topic: 4. Presenting Data Visually
4.3. Apply conditional formatting
4.3.2. Allow more than one rule to be true

Excel: Overtime (Page 1) 19 20 21 22 23 24 25 26 27 28 29 30 **31** 32 33 34 35 36 37 38

Home -> Conditional Formatting -> Data Bars

Data Bars

The following exercises demonstrate the various **Conditional Formatting** available in Microsoft Excel. All of these examples will be applied to the H Column, the one with sales data.

Try It: Format with Data Bars
Select Column H.
Go to **Home -> Conditional Formatting**.
Select: **Data Bars**.

What Do You See? The cell with the highest value is 100% filled in. The other cells are filled with less than 100%. Excel calculates the percentage.

After each trial, select the H Column.
Go to **Conditional Formatting -> Clear Rules -> From the Selected Cells.**

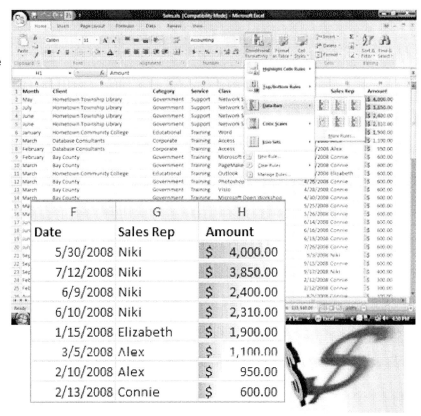

Microsoft Excel 2007 Exam 77-602 Topic: 4. Presenting Data Visually
4.3. Apply conditional formatting: Data bars

Excel: Overtime (Page 1) 19 20 21 22 23 24 25 26 27 28 29 30 31 32 33 34 35 36 37 38

Home -> Conditional Formatting -> Color Scales

Color Scales
Try It: Format with Color Scales
Select Column H.
Go to **Home -> Conditional Formatting**.
Select **Color Scales**: you will have a short set of built-in options.

What Do You See? The Color Scales use two colors to indicate the Conditional Formatting. In this example, the highest value is red and the lowest is yellow. Microsoft Excel compares each amount and formats the cells accordingly.

After each trial, select the H Column.
Go to **Conditional Formatting -> Clear Rules -> From the Selected Cells.**

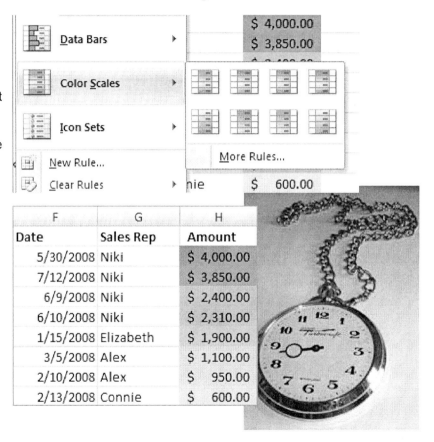

Microsoft Excel 2007 Exam 77-602 Topic: 4. Presenting Data Visually
4.3. Apply conditional formatting: Color Scales

Excel: Overtime (Page 1) 19 20 21 22 23 24 25 26 27 28 29 30 31 32 33 34 35 36 37 38

Home -> Conditional Formatting -> Icon Sets

Icon Sets

Try It: Format with Icon Set
Select Column H.
Go to **Home -> Conditional Formatting.**
Select **Icon Sets:** you will have a short set of built-in options.

What Do You See? The Icon Sets use a three- color scale to show the Formatting.
By default, the percents are:
Highest: >67% (Greater Than).
Middle: <67 and >33%. (Between).
Low: <33% (Less Than).

Excel: Overtime (Page 1) 19 20 21 22 23 24 25 26 27 28 29 30 31 32 33 34 35 36 37 38

Home -> Conditional Formatting -> Manage Rules

Edit the Rules
Is there a way to edit and manage the **Conditional Formatting Rules**? Certainly. ;-)

Try It: Use the Rules Manager
Go to **Home -> Conditional Formatting**. Click on **Manage Rules**.

What Do You See? The Rules Manager should display a list of the Conditional Formatting. You can add, delete, and edit the Rules here.

Memo to Self: The default view is **Current Selection**. You can change the selection to show all of the formatting rules for the entire spreadsheet, or on a different sheet.

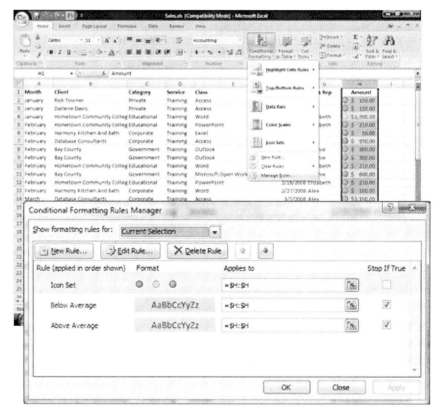

Microsoft Excel 2007 Exam 77-602 Topic: 4. Presenting Data Visually
4.3. Apply conditional formatting
4.3.1. Use the Conditional Formatting Rules Manager: Edit an existing conditional formatting rule

Excel: Overtime (Page 1) 19 20 21 22 23 24 25 26 27 28 29 30 31 32 33 34 35 36 37 38

Home -> Conditional Formatting -> Manage Rules

Bend the Rules

A couple of pages ago you created a Conditional Format that used Icon Sets. The little icon formatting was applied to the cells based on their values. Here are the steps to edit the values in that Rule.

Try It: Edit the Rule
Go to **Home -> Conditional Formatting**
Click on **Manage Rules**.
Select the Icon Set Rule.
Go to **Edit Rule**.

What Do You See? The **Rule Type** is a list of options. You can use a rule to identify a cell by the values. The values can be based on the average or the rank. You can also use a formula in a rule.

The **Rule Description** is a tool for defining the values. You can select the formatting **Style** and type in your own **Value** if you wish.

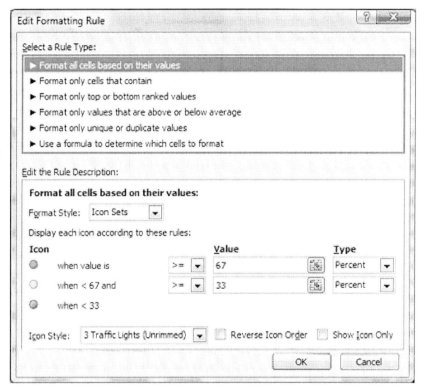

Excel: Overtime (Page 1) 19 20 21 22 23 24 25 26 27 28 29 30 31 32 33 34 35 36 37 38

Home -> Editing ->Find and Select

Find Special Cells

You can use the **Find** option to locate the cells which use Conditional Formatting or Data Validation.

The Find button is in the Editing group on the Home Ribbon. The Find icon may look different in Word and Excel (big or small), but in both programs it is still on the right side of the Ribbon.

Try This: Use the Find Options
Go to **Home ->Editing.**
Select **Find and Select**.
Click on Data Validation or Conditional Formatting and test the Find options.

Will this option find the Data Validation or Conditional Formatting on a different spreadsheet in the same workbook?

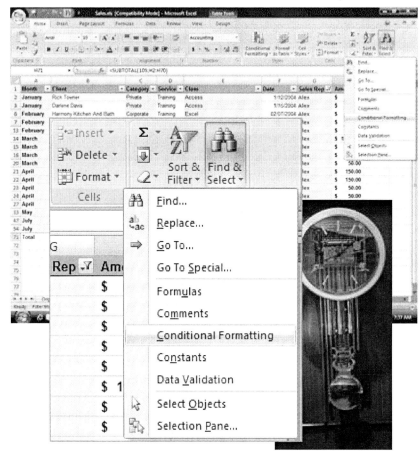

Microsoft Excel 2007 Exam 77-602 Topic: 4. Presenting Data Visually
4.3. Apply conditional formatting

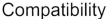

Excel: Overtime (Page 1) 19 20 21 22 23 24 25 26 27 28 29 30 31 32 33 34 35 36 **37** 38

Office -> Prepare -> Compatibility Checker

Compatibility

Microsoft Excel 2007 offers many rich, creative features. However, many of these options are not supported in earlier versions of Excel. For example, Excel 2000 cannot "see" the Icon Set formatting because the Icon programming simply isn't there in the old Excel software. Previous versions of Excel used Lists, which Excel 2007 doesn't use. That capability is now available as Tables.

You may get a **Compatibility** message when you save your Excel 2007 work. You can also check the compatibility with the Office menu.

Try It: Check Compatibility
Go to **Office ->Prepare**.
Select **Compatibility Checker**.

What Do You See? The summary will indicate which new features are not compatible with older versions of Excel. If you want to keep the old functions, you can Copy to a New Sheet, then **Continue**.

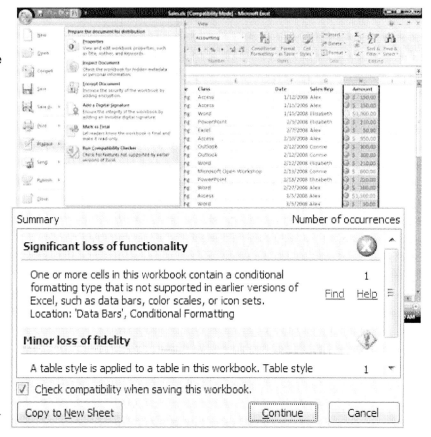

Excel: Overtime (Page 1) 19 20 21 22 23 24 25 26 27 28 29 30 31 32 33 34 35 36 37 38

DONE

Formatting Data

This demonstration illustrated dynamic formatting that changes based on whatever is typed in the cells. Data Validation and Conditional Formatting were used to create interactive forms that are efficient and user friendly.

Well done, that took some extra thought. You can have two cookies.

Test Yourself

1. Data Validation includes options to restrict the data in a cell by which criteria? (Select all correct answers.)
a. Decimals
b. Whole Numbers
c. Dates
d. Times
e. Items from a Reference List
   Tip: Intermediate Excel, (page 49)

2. Which of the following are true about Data Validation options? (Select all correct answers.)
a. An input message can be set to guide users
b. An error message can be set to alert users to a mistake
c. An error message is set by default by Excel
   Tip: Intermediate Excel, (page 50-1)

3. Graphics can be added to Excel Spreadsheets.
a. TRUE
b. FALSE
   Tip: Intermediate Excel, (page 56)

4. Which command combines several cells into one large separates?
a. Home-> Alignment-> Merge and Center
b. View-> Alignment-> Merge Cells
c. Home-> Cells-> Combine
d. Format-> Alignment-> Merge Cells
   Tip: Intermediate Excel, (page 58)

5. Which of the following is a type of Conditional Formatting?
(Select all correct answers.)
a. Color Scales
b. Highlight
c. Color Bars
d. Icon Sets
   Tip: Intermediate Excel, page (65-9)

Page 1 2 3 4 5 6 7 8 9 10 11 12 13 14 15 16 17 18 19 20 21 22 23 24 25 26 27 28 29

Sound Advice

Legs, Eggs and Pigs in a Basket

Click Here to Get Started
Sample Files

Intermediate Excel

Lesson Objectives: Learn how to create a sales summary that combines information from several spreadsheets. This lesson will demonstrate how to create formulas that reference data from other worksheets, use absolute and relative cell references, and how to troubleshoot your equations. In this lesson you will:

Practice formatting data and content page 1

Demonstrate how to fill a series with AutoFill page 7

Learn how to calculate Revenue with a formula page 10

Practice the steps needed to summarize the data page 15

Identify Relative References page 18

Identify errors and troubleshoot the equation page 21

Learn how to create and use Absolute References page 22

Practice using the Auditing Toolbar page 23

Compare What If Scenarios and Goal Seek page 25

Excel: Legs, Eggs, Pigs Page 1 2 3 4 5 6 7 8 9 10 11 12 13 14 15 16 17 18 19 20 21 22 23 24 25 26 27 28 29

Calculating Sales

Our little online farmer's market is really growing. Supply can hardly keep up with demand at Charlotte's Web Site. Is there some way we can forecast what we might sell, say Wednesday two weeks from now? Yes, there is. We can set up a spreadsheet that will calculate our sales on any given day. **Start** the **Program** Microsoft **Excel**.

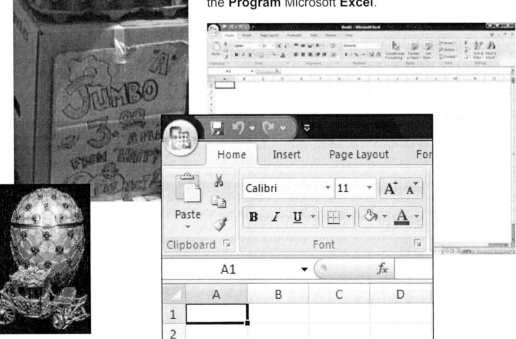

What do you see from the top of the screen? Is there a blue **Title Bar** that says Microsoft Excel? Yes.

Is there a **Home** Ribbon with the Clipboard, Font, Alignment and Number Groups? Yes.

If your screen looks similar to the example on this page, then you are ready to get started.

Excel: Legs, Eggs, Pigs Page 1 2 3 4 5 6 7 8 9 10 11 12 13 14 15 16 17 18 19 20 21 22 23 24 25 26 27 28 29

Home -> Font -> Bold

Enter The Data
1. There are only three parts to a spreadsheet: **labels**, **data**, and **formulas**. Start with the labels.

Try it:
Click on cell A1 and type: Date
In cell B1 type: Product
In cell C1 type: Net
In cell D1 type: Quantity
In cell E1 type: Revenue

If these are **labels**--and they are-- select Row 1 and make them **Bold**.

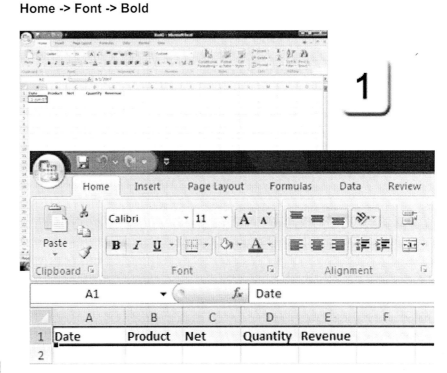

Microsoft Excel 2007 Exam 77-602 Topic: 2. Formatting Data and Content
2.3. Format cells and cell content: Format cells using other methods

Excel: Legs, Eggs, Pigs Page 1 2 3 4 5 6 7 8 9 10 11 12 13 14 15 16 17 18 19 20 21 22 23 24 25 26 27 28 29

Home -> Number

Format Date and Time

2. Select cell A2 and enter the date. Please type: 6/1/07. What do you see? The default **format** for date changed your display to 1-Jun-07.

Look on the formula. In the sample screen on this page, it reads exactly the way it was typed.

The FORMAT in cell A1 displays d/m/yy. This is formatting, the same way text can be formatted big, bold and colorful. The DATA is what you see in the Formula bar.

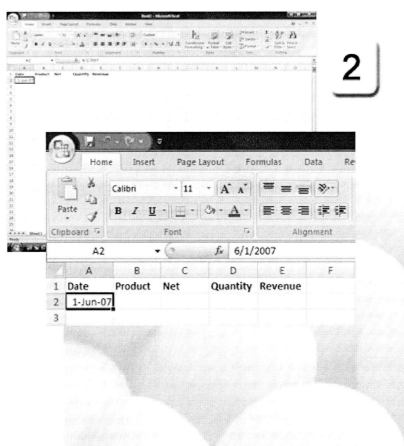

Microsoft Excel 2007 Exam 77-602 Topic: 2. Formatting Data and Content
2.3. Format cells and cell content: Format the values in a cell as a date

Excel: Legs, Eggs, Pigs Page 1 2 3 4 5 6 7 8 9 10 11 12 13 14 15 16 17 18 19 20 21 22 23 24 25 26 27 28 29

Home -> Number ->Format Cells

Format Date and Time

3. You can format this column for date and time. Go to the **Home** Ribbon and click on the **Number** group.

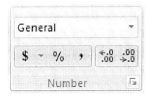

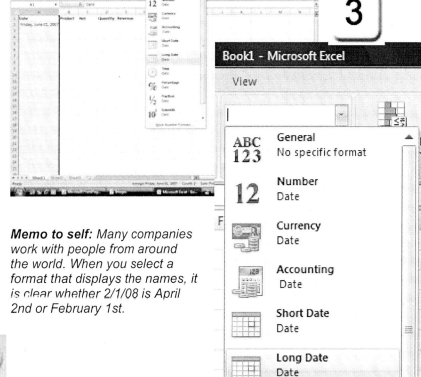

Try it: Use the combo box to change the **General** format. The drop-down box shows some options. The **Long Date** format spells out the months.

Memo to self: *Many companies work with people from around the world. When you select a format that displays the names, it is clear whether 2/1/08 is April 2nd or February 1st.*

Microsoft Excel 2007 Exam 77-602 Topic: 2. Formatting Data and Content
2.3. Format cells and cell content: Format the values in a cell as a date

Excel: Legs, Eggs, Pigs Page 1 2 3 4 5 6 7 8 9 10 11 12 13 14 15 16 17 18 19 20 21 22 23 24 25 26 27 28 29

Home -> Number ->Format Cells

Format Cells

4. There are additional formats for the Number Group. Go to the option **arrow** in the lower right corner.

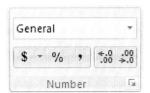

You should see several pages of formatting. On the **Number** page, select the **Date** category and select the Medium Type: March 14, 2001.

Memo to Self: You can create a Custom format if you wish.

Microsoft Excel 2007 Exam 77-602 Topic: 2. Formatting Data and Content
2.3. Format cells and cell content: Format the values in a cell as a date

Excel: Legs, Eggs, Pigs Page 1 2 3 4 5 6 7 8 9 10 11 12 13 14 15 16 17 18 19 20 21 22 23 24 25 26 27 28 29

AutoFill the Data

5. Now, one way to put in the dates for the next month would be to type them all. Do you want to know a faster way?

Click once on cell A2 to select it. Do you see the little **AutoFill** handle at the bottom of the cell? It's a little black square. Move the mouse over the handle. When the cursor changes from a fat white plus sign to a thin black cross, hold your left mouse button down. Drag down to cell A20 and release.

Charlotte's Website sells eggs, chicken wings, and little sausages wrapped in pastry. Eggs, Legs, and Pigs in a Basket. Under product type, "eggs."

AutoFill works here, too. Click once on Eggs to select it. Double click the handle and the word Eggs fills all the way down. Look, it goes down as far as the column to the right. So don't worry, it won't go on beyond zebra.

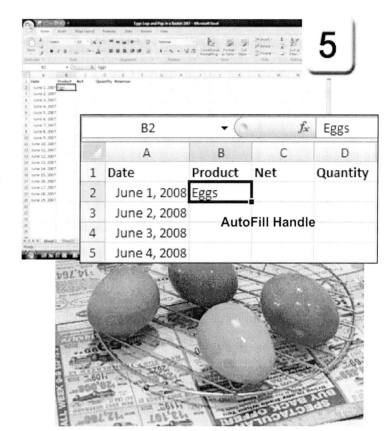

Microsoft Excel 2007 Exam 77-602 Topic: 1. Creating and Manipulating Data
1.1.1. Fill a series   1.1.2. Copy a series

Excel: Legs, Eggs, Pigs Page 1 2 3 4 5 6 7 8 9 10 11 12 13 14 15 16 17 18 19 20 21 22 23 24 25 26 27 28 29

Home ->Number

Format Currency

6. Now, the net profit on a dozen eggs is $1.50 after shipping and handling. Hey! These are gourmet eggs, sold on the Internet. Under Net, type "1.50."

You don't need to type the dollar sign. We're going to format the column. Select the Net column by clicking on the C column header. Now, click on **currency** (dollar sign) in the **Number** group. Excel formats the column so the dollars and cents line up.

AutoFill the $1.50 in cell C2 by selecting the cell and double clicking on the little black handle.

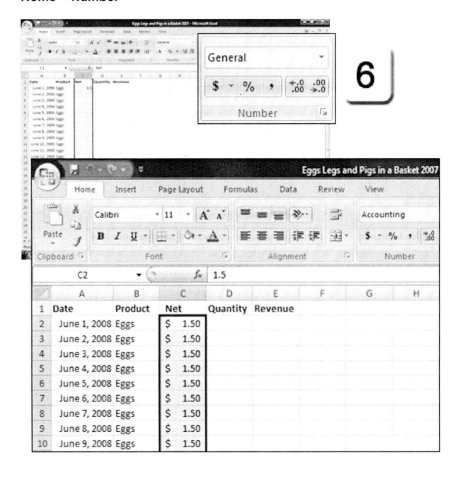

Microsoft Excel 2007 Exam 77-602 Topic: 2. Formatting Data and Content
2.3. Format cells and cell content   2.3.1. Apply number formats

Excel: Legs, Eggs, Pigs Page 1 2 3 4 5 6 7 8 9 10 11 12 13 14 15 16 17 18 19 20 21 22 23 24 25 26 27 28 29

AutoFill A Series

Our little web site sells about 100 egg baskets a day. Suppose we sold another 5 baskets every day. What would our revenue be?

Try it: In cell D2 type 100.
In cell D3 type 105.

AutoFill. Select both the 100 and the 105. Now you have selected enough of a range for Excel to recognize that this is a series, incrementing by 5s.

Double click the AutoFill handle and the series will fill down to match the C column on the left.

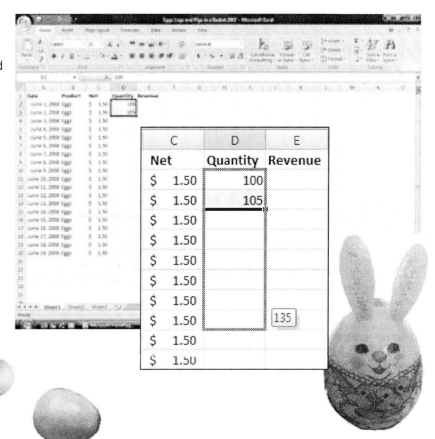

Microsoft Excel 2007 Exam 77-602 Topic: 1. Creating and Manipulating Data
1.1.1. Fill a series 1.1.2. Copy a series

Excel: Legs, Eggs, Pigs Page 1 2 3 4 5 6 7 8 9 <u>10</u> 11 12 13 14 15 16 17 18 19 20 21 22 23 24 25 26 27 28 29

Calculate Revenue

Labels and Data. Now all we need are the formulas. Select cell E2 and type in the equal sign. All good equations begin with "=."

Think about your high school math. Revenue equals net times quantity. Click on cell C2—the Net of $1.50. The times symbol is made with the asterisk key. Now click on cell D2—the quantity. The equation is =C2*D2.

Hit the Enter key. Excel won't calculate until you leave the cell. There it is. $1.50 times 100 equals $150.00. OK, go to cell E2 and double click on the AutoFill handle to fill in the equations.

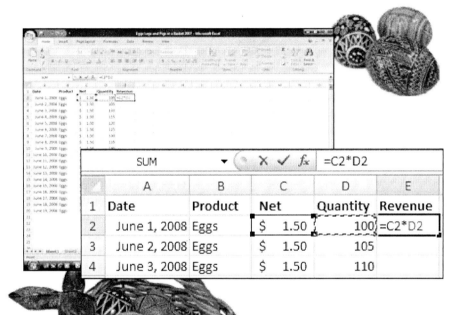

Excel: Legs, Eggs, Pigs Page 1 2 3 4 5 6 7 8 9 10 11 12 13 14 15 16 17 18 19 20 21 22 23 24 25 26 27 28 29

Home ->Editing ->AutoSum

Subtotals

What is the sum of all the products sold? Here are the steps to create a total for the daily sales.

Try it: AutoSum the Revenue
Select cell E21
Go to the **Home Ribbon**
Click on the **AutoSum Function**

Excel will enter the following equation:
=SUM(E2:E20)

Click ENTER on your keyboard and Excel will calculate the Sum.

18	June 17, 2008	Eggs	$	1.50	180	$	270.00
19	June 18, 2008	Eggs	$	1.50	185	$	277.50
20	June 19, 2008	Eggs	$	1.50	190	$	285.00
21						=SUM(E2:E20)	

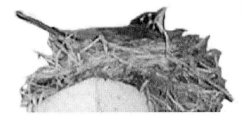

Microsoft Excel 2007 Exam 77-602 Topic: 3. Creating and Modifying Formulas
3.2. Summarize data using a formula   3.2.1.   Use SUM, COUNT, COUNTA, AVERAGE, MIN, and MAX

Excel: Legs, Eggs, Pigs Page 1 2 3 4 5 6 7 8 9 10 11 **12** 13 14 15 16 17 18 19 20 21 22 23 24 25 26 27 28 29

Copy the Spreadsheet
OK, that takes care of the eggs. Let's name the spreadsheet and move on.

Try it: Rename the Spreadsheet
Double click the Sheet1 tab
Type, "Eggs."

Copy the Spreadsheet
Right mouse click the "Eggs" label
Select **Move or Copy**
Check **Create A Copy**

Say OK and there is a new sheet called Eggs (2). This method copied everything: labels, data, and formulas. Saves time, doesn't it? Now all we have to do is change a few variables

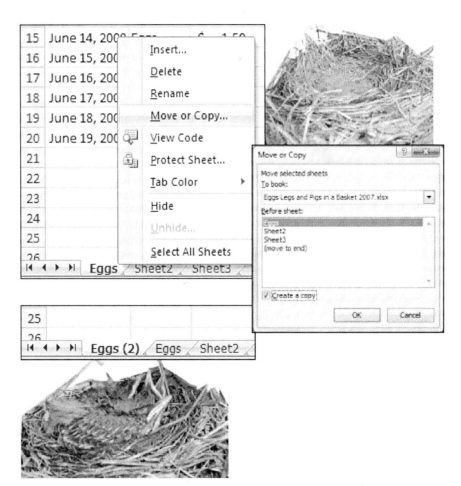

Microsoft Excel 2007 Exam 77-602 Topic: 1. Creating and Manipulating Data:
1.5. Manage worksheets   1.5.1. Copy worksheets

Excel: Legs, Eggs, Pigs Page 1 2 3 4 5 6 7 8 9 10 11 12 13 14 15 16 17 18 19 20 21 22 23 24 25 26 27 28 29

Change the Variables

Rename the Spreadsheet
Double click on the tab for Eggs (2).
Rename this spreadsheet: Legs.

Update the product data
Go to cell B2 and change Eggs to Legs.
AutoFill cell B2 to the bottom of Column B.

Change the Net
The net for chicken legs is 3.25.
Enter that amount and AutoFill.

Change the Quantity
For quantity we will start at 50 and add 5 per day. Type 50 in cell D2.
Type 55 in D3
Select D2 and D3 and AutoFill.

The work on the Pig's page is similar. Rename the spreadsheet to Pigs in a Basket. Change the product to Pigs. The Net is 4.75. and the quantity starts at 200 but only adds one more per day.

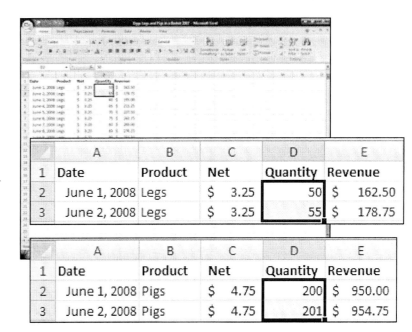

Excel: Legs, Eggs, Pigs Page 1 2 3 4 5 6 7 8 9 10 11 12 13 14 15 16 17 18 19 20 21 22 23 24 25 26 27 28 29

Home ->Cells ->Format ->AutoFit Column Width

AutoFit the Columns
Now, hold on. Hold on. The E column may display #####. Why are there pound signs? That's Excel's way of telling us that the column is too narrow.

Try it: Format the Column Width
Select Column E
Select **Format** on the Home Ribbon
Select Format

Here's another way:
Place your cursor between the E and F column header—right on the line. Your cursor will change from a white plus sign to a black double-headed arrow. Double click and the column will be made as wide as it needs to be.

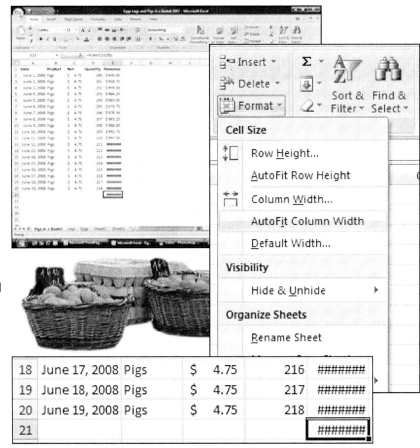

Microsoft Excel 2007 Exam 77-602 Topic: 2. Formatting Data and Content
2.2. Insert and modify rows and columns: Using AutoFit

Excel: Legs, Eggs, Pigs Page 1 2 3 4 5 6 7 8 9 10 11 12 13 14 15 16 17 18 19 20 21 22 23 24 25 26 27 28 29

Home ->Cells -> Delete ->Delete Sheet Columns

Create a Summary Sheet

1. Make a copy of the "Eggs" spreadsheet. Double click the tab and rename it "Summary."

2. **Change** the product from Eggs to All Products and **AutoFill** cell B2.

3. Select column C and D, and delete them. We don't need them. Go to **Home ->Cells**.
Select **Delete ->Delete Sheet Columns**.

What do you see? The equations in the Revenue column now read: #REF!. The equation doesn't have the data for Net and Quantity anymore. It has nothing to reference, hence the error message. Go ahead: select those cells and hit the delete key to clear the busted equation.

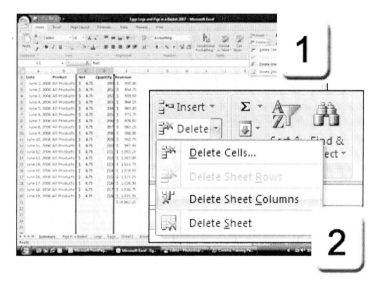

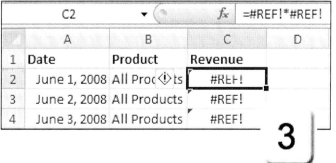

Microsoft Excel 2007 Exam 77-602 Topic: 3. Creating and Modifying Formulas:
3.1. Reference data in formulas
3.1.1. Create formulas that use absolute and relative cell references: Troubleshoot a formula

Excel: Legs, Eggs, Pigs Page 1 2 3 4 5 6 7 8 9 10 11 12 13 14 15 16 17 18 19 20 21 22 23 24 25 26 27 28 29

The Grand Total
This equation will link the Summary sheet to each of the product sheets: Legs, Eggs, and Pigs in a Basket.

Try it: Create reference links
1. Start on the **Summary** Sheet
Select Cell C2 type: =.
All good equations begin with "="

2. Go to the **Legs** sheet
Select cell E2.
Type + (to add the next variable)

3. Go to the **Eggs** sheet
Select cell E2.
Type + (to add the next variable)

Panel 1 — SUM fx =

	A	B	C	D
1	Date	Product	Revenue	
2	June 1, 2008	All Products	=	
3	June 2, 2008	All Products		
4	June 3, 2008	All Products		

Panel 2 — SUM fx =Legs!E2

	A	B	C	D	E
1	Date	Product	Net	Quantity	Revenue
2	June 1, 2008	Legs	$ 3.25	50	$ 162.50
3	June 2, 2008	Legs	$ 3.25	55	$ 178.75
4	June 3, 2008	Legs	$ 3.25	60	$ 195.00

Panel 3 — SUM fx =Legs!E2+Eggs!E2

	A	B	C	D	E
1	Date	Product	Net	Quantity	Revenue
2	June 1, 2008	Eggs	$ 1.50	100	$ 150.00
3	June 2, 2008	Eggs	$ 1.50	105	$ 157.50
4	June 3, 2008	Eggs	$ 1.50	110	$ 165.00

Microsoft Excel 2007 Exam 77-602 Topic: 3.1.2. Create formulas that reference data from other worksheets
3.1.2. Create formulas that reference data from other worksheets or workbooks

Excel: Legs, Eggs and Pigs Page 1 2 3 4 5 6 7 8 9 10 11 12 13 14 15 16 17 18 19 20 21 22 23 24 25 26 27 28 29

The Grand Total

4. Lastly, go to the **Pigs** sheet
Select cell E2
Hit Enter on your keyboard so Excel will calculate.

5. You will be taken back to cell C2 in the Summary spreadsheet.

6. **AutoFill** this equation and we've got a formula that adds all of the sales for each product. Well done.

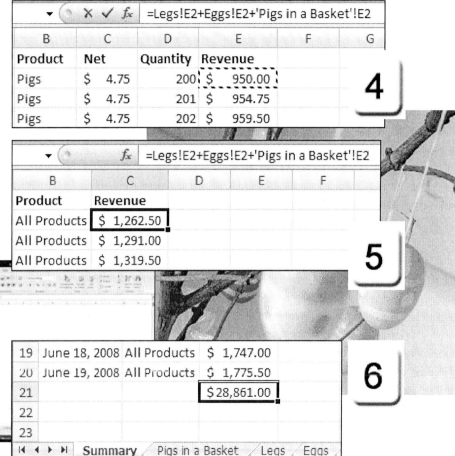

Page 91 of 115

Excel: Legs, Eggs, Pigs Page 1 2 3 4 5 6 7 8 9 10 11 12 13 14 15 16 17 18 19 20 21 22 23 24 25 26 27 28 29

Cell References

When you created the sales spreadsheet, you selected a **range of cells** to **AutoFill**. If we put 100 for the quantity in the first cell and 105 in the next, Excel fills down the series and add 5 more to each cell. If you wanted a different forecast you could enter new quantities in the first and second cells and use the AutoFill again.

This could get old very fast. It is also a rather inflexible method of changing the data. There is a better method: **reference cells**.

Reference cells set up one place to enter the data. All of the other equations and spreadsheets that depend on that data look it up in the reference cells.

Excel: Legs, Eggs, Pigs Page 1 2 3 4 5 6 7 8 9 10 11 12 13 14 15 16 17 18 19 20 21 22 23 24 25 26 27 28 29

Home->Cells ->Insert

Create a Reference Cell

Try it: Open the spreadsheet we created in Eggs, Legs, and Pigs in a Basket.

Select row 2 and **Insert** three more **Rows**. This will give us room to work.

In cell A1 type the label, "Initial."

In cell A2 type, "Increment." Now, enter 100 in cell B1 and 5 in cell B2.

Calculate the sales: Cell D6 equals whatever we sold on the first day, plus the quantity we are forecasting for each day's sales increase.

That's your next task...

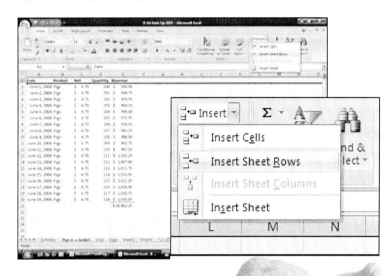

Excel: Legs, Eggs, Pigs Page 1 2 3 4 5 6 7 8 9 10 11 12 13 14 15 16 17 18 19 20 21 22 23 24 25 26 27 28 29

Create a Reference Cell

Read across your spreadsheet: cell D5 is the initial quantity sold. Cell D6 should be whatever we sold on the first day, plus the quantity we are forecasting for each day's sales increment.

Try it: Create a Reference Cell
Select D5 and enter this equation:
=B1

Select D6, and enter this equation:
=D5 +B2

Test your equation: change the Initial number in cell B2 from 100 to 90. What do you see in cell D5 or D6?

Now, double click the AutoFill handle to fill down the series.

Troubleshoot the Equation

What do you see? Where did **#Value!** come from? Double click on cell D8. Excel highlights the cells in this equation.
=D7+B4

How did that happen? Part of the equation is still correct. It says start with yesterday's sales—the cell above me—and add to it.

Most equations automatically use **Relative References**. A Relative Reference adjusts the the cell references when you copy, or fill it down. For example, **=D7+B4** becomes **=D8+B5** in the next row down.

By the time you get to cell D8, this equation is adding a number to a label. Hence, the error #Value!

	A	B	C	D	E
			SUM	▼ X ✓ fx	=D7+B4
1	Initial	100			
2	Increment	5			
3					
4	Date	Product	Net	Quantity	Revenue
5	June 1, 2008	Pigs	$ 4.75	100	$ 475.00
6	June 2, 2008	Pigs	$ 4.75	105	$ 498.75
7	June 3, 2008	Pigs	$ 4.75	105	$ 498.75
8	June 4, 2008	Pigs	$ 4.75	=D7+B4	#VALUE!
9	June 5, 2008	Pigs	$ 4.75	#VALUE!	#VALUE!
10	June 6, 2008	Pigs	$ 4.75	#VALUE!	#VALUE!

Microsoft Excel 2007 Exam 77-602 Topic: 3. Creating and Modifying Formulas:
3.1. Reference data in formulas
3.1.1. Create formulas that use absolute and relative cell references: Troubleshoot a formula

Excel: Legs, Eggs, Pigs Page 1 2 3 4 5 6 7 8 9 10 11 12 13 14 15 16 17 18 19 20 21 22 23 24 25 26 27 28 29

Absolute References
When you need to work with one particular cell you need an **Absolute Reference**.

Try it: Select cell D6. The Formula bar shows =D5+B2.

Now, click on B2 in the Formula bar. When you hit the **F4 function** on the top row of the keyboard, the cell reference becomes $B$2. This means "go to B2 only, and no place else," to get the data.

Select cell D6 and **AutoFill** the revised equation to the rest of the rows.

A **Mixed Reference** uses Absolute and Relative cell references.

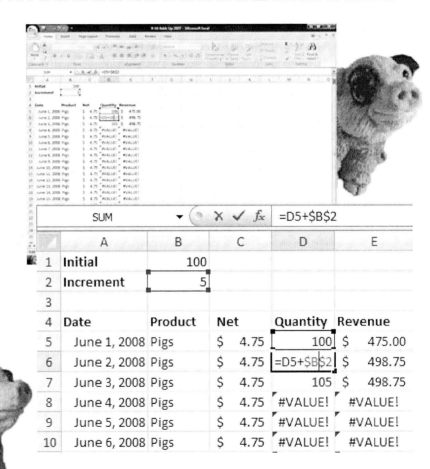

Microsoft Excel 2007 Exam 77-602 Topic: 3. Creating and Modifying Formulas
3.1. Reference data in formulas
3.1.1. Create formulas that use absolute and relative cell references :that maintains its reference point when copied

Excel: Legs, Eggs, Pigs Page 1 2 3 4 5 6 7 8 9 10 11 12 13 14 15 16 17 18 19 20 21 22 23 24 25 26 27 28 29

Formulas -> Formula Auditing

Check Your Work

In the previous pages, we double clicked on a cell to **trace** the references in the equation. There is a more sophisticated method for examining your work.

On the **Formulas** Ribbon, go to **Formula Auditing**. There are some useful tools available:

Trace Precedents
Finds the cells that have an affect on the cell you selected

Trace Dependents
Locates the formulas that use this cell in their equations

Remove the Arrows

Microsoft Excel 2007 Exam 77-602 Topic: 3. Creating and Modifying Formulas:
3.1. Reference data in formulas
3.1.1. Create formulas that use absolute and relative cell references: Troubleshoot a formula

Excel: Legs, Eggs, Pigs Page 1 2 3 4 5 6 7 8 9 10 11 12 13 14 15 16 17 18 19 20 21 22 23 24 25 26 27 28 29

Formulas -> Formula Auditing

Auditing

Try it: Select cell E7 and click **Trace Precedents**. You will see an arrow from cell C7 and cell D7.

Click on **Trace Precedents** again. you can see where D7 gets its data.

Remove All Arrows and try another audit experiment.

Select B2 and **Trace Dependents.** All of the equations in the D column depend on the data in B2 in order to calculate correctly.

Click Trace Dependents a second time, and you will see a call out to another worksheet.

Memo to Self: the Revenue for each product is added together on the Summary sheet. The equation was:

=Legs!E2+Eggs!E2+'Pigs in a Basket'!E5

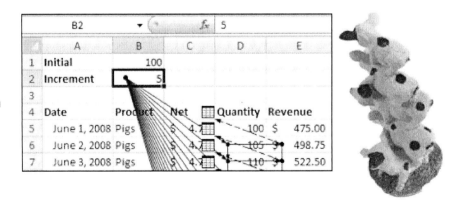

Microsoft Excel 2007 Exam 77-602 Topic: 3. Creating and Modifying Formulas:
3.1. Reference data in formulas
3.1.1. Create formulas that use absolute and relative cell references: Troubleshoot a formula

Excel: Legs, Eggs, Pigs Page 1 2 3 4 5 6 7 8 9 10 11 12 13 14 15 16 17 18 19 20 21 22 23 24 25 26 27 28 29

Data -> Data Tools ->What-If Analysis

What If Scenarios

Suppose you wanted to look up the best and worst cases for your sales projections. You could type in a robust initial sales number and a vigorous daily increment for the best case. Then, you could type in more conservative estimates for the worst case.

Microsoft Excel has a way to save that information: What-If Analysis.

Try it: Use the What-If Analysis

1. Select cell B1. Now, define the scenario by going to **Data Tools** on the Data Ribbon and clicking **What-If Analysis**.

2. When the **Scenario Manager** opens, please look for the Add button on the right.

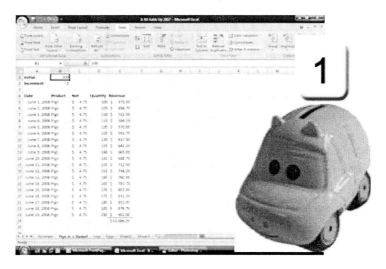

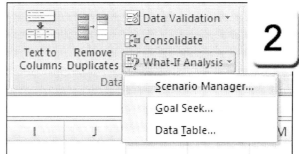

Excel: Legs, Eggs, Pigs Page 1 2 3 4 5 6 7 8 9 10 11 12 13 14 15 16 17 18 19 20 21 22 23 24 25 26 27 28 29

Data -> Data Tools ->What-If Analysis ->Scenario Manger

Scenario Manager

3. **Name** the scenario: Best Case

4. **Type:** B1 where it says **Changing Cells**. Remember, B1 is the cell for Initial Sales.

5. When you are done editing the scenario, you will be prompted to **Enter a value**. Please type in 100. After you click on OK you will return to the Scenario Manager.

Add another scenario for Worst Case and change the quantity for B2 from 100 to 50.

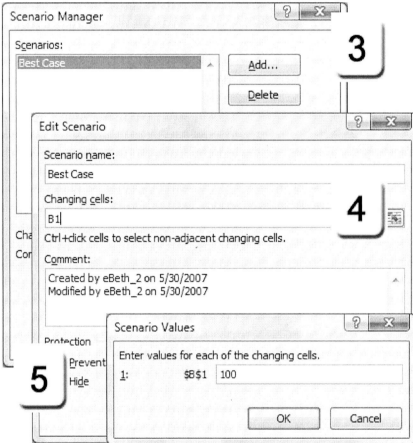

Excel: Legs, Eggs, Pigs Page 1 2 3 4 5 6 7 8 9 10 11 12 13 14 15 16 17 18 19 20 21 22 23 24 25 26 27 28 29

Data -> Data Tools -> What-If Analysis -> Scenario Manager

Add to the Scenario
What if you wanted to change both the Initial goal and the Increment?

1. **Edit** the Best Case scenario

2. Select which cells to use in the scenario by clicking the red, white and blue selection button by the **Changing Cells** option.

Click on cell B1, then hold the Control key on the keyboard as you select B2. Close the window.

3. When you click on OK, the **Scenario Manager** will prompt you to enter a new value for each reference cell. Type 200 for cell B1 and 15 for cell B2

Create another Scenario of the Worst Case. Let $B$1 equal 30 and $B$2 equal 3.

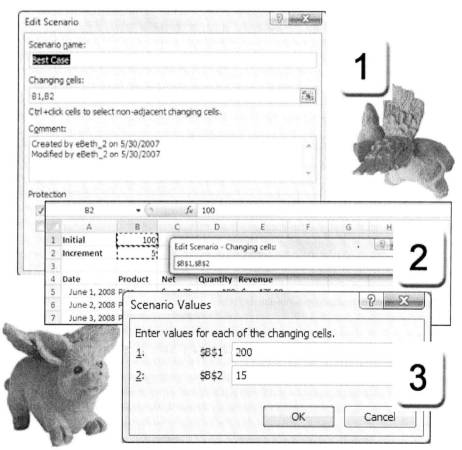

Excel: Legs, Eggs, Pigs Page 1 2 3 4 5 6 7 8 9 10 11 12 13 14 15 16 17 18 19 20 21 22 23 24 25 26 27 28 29

Data -> Data Tools ->What-If Analysis -> Goal Seek

Goal Seek

There is another way to create a forecast. All of our formulas calculated total revenue from the initial quantity, added more sales each day and multiplied the quantity by the net.

What if you want to work backwards and ask, "How many Pigs in a Basket do I have to sell to get $40,000?"

Try it: Use Goal Seek
1. Start by going to the total in cell E24 on the Eggs sheet. Now, go to **Data Tools** on the **Data** Ribbon and choose **Goal Seek**.

2. You will see a dialog box that asks you to fill in the value for our set cell, E24, by changing another cell in our spreadsheet.

Excel: Legs, Eggs, Pigs Page 1 2 3 4 5 6 7 8 9 10 11 12 13 14 15 16 17 18 19 20 21 22 23 24 25 26 27 28 29

DONE

Data -> Data Tools ->What-If Analysis -> Goal Seek

Goal Seek

3. For this experiment, use the red, white and blue button to pick our increment in cell B2. Close the window.

4. After you say OK, you will see the **Goal Seek** status. Look at the result in cell B2. In order to meet our goal of $40,000, Charlotte's Website needs to sell an additional 27 Pigs in a Blanket each day. That's ambitious!

Tools of the Trade
All of the work we have done today used options we found in the Tools menu. Auditing, Scenarios, and Goal Seeking are all ways to make a spreadsheet more interactive.

Well, you done good. You get the cookie.
<grin>

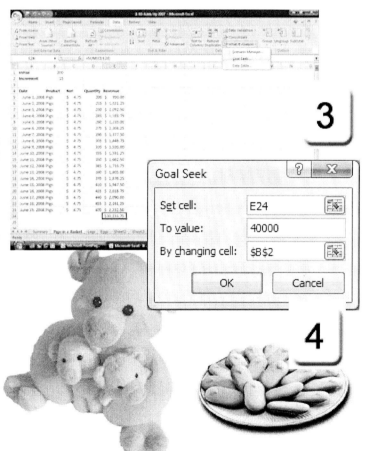

Test Yourself

1. The Number group on the Home menu has options for formatting the date and time. The Long Date format spells out the months.
a. TRUE
b. FALSE
 Tip: Intermediate Excel, p 79

2. What does the error message ###### mean?
a. One or more of the cells in the formula contains text instead of numbers
b. The formula is too large to fit in the cell
c. The formula includes an absolute reference
 Tip: Intermediate Excel, p 88

3. $B$2 is an example of what kind of cell reference?.
a. Relative
b. Absolute
 Tip: Intermediate Excel, p. 96

4. What If Scenarios allow you to save several versions of your data with the Scenarios Manager.
a. TRUE
b. FALSE
 Tip: Intermediate Excel, p 99

5. Which of the following are true about What-If Analysis? (Select all correct responses)
a. You will be prompted to name the scenario
b. You need to indicate which cell(s) will change
c. You can change more than one cell
 Tip: Intermediate Excel, p 100

6. Which of the following are true about Goal Seek? (Select all correct answers)
a. You will be prompted to select a cell to change
b. You need to indicate the new value
c. You can save several different Goals using Goal Seek
 Tip: Intermediate Excel, p 103

Assessment

Practice Activities
Complete the Online Quiz for this Level.

Skill Test: Download the instructions.
Submit the Skill Test to your instructor

Intermediate Microsoft® Excel: Practice 1

RUBRIC

0	3	5	8	10
Less than 25% of items completed correctly.	More than 25% of items completed correctly	More than 50% of items completed correctly	More than 75% of items completed correctly	All items completed correctly

Each step to complete is considered a single item, even if it is part of a larger string of steps.

Objectives:
The Learner will be able to
1. Apply **Currency** formatting to cells in Excel
2. Use the **AutoSum** tool to add cells
3. Use the **Function** tool to calculate the average of a range of cells
4. Insert an **IF** function
5. Rename spreadsheet

Working with Equations

Enter the labels and format the labels big, bold, and centered

- In Cell A1 type: First Name
- In Cell B1 type: Last Name
- In Cell C1 type: Department
- In Cell D1 type: Salary

Add the data
Type at least five sample records. For example:

	A	B	C	D
1	**First Name**	**Last Name**	**Department**	**Salary**
2	Deeter	Poohbah	Training	$34,000

Format the Columns
Select column D and use the Currency tool

Creating Equations
Select Cell D7 and use AutoSum to add up the SUM of the Salaries in the D Column.
Select Cell D8 and use Insert Function to calculate the AVERAGE of the Salaries.

Using IF functions
This activity compares the employee's salary with the Average in Cell D8.
Insert the label "Compare" into cell E1
In E2, insert the IF function.
  In the Wizard, enter the following information
    Logical Test: D2>D8
    Value_if_true: "Above"
    Value_if_false: "Below"
Use the Insert Function wizard to put the correct formula for the remaining cells.

Save the spreadsheet and name it: Excel Intermediate Practice 1

Intermediate Microsoft® Excel: Practice 2

Objectives:

The Learner will be able to:
1. Explain what labels are
2. Sort Excel data by using the labels in the header row
3. Create a Custom Sort
4. Modify the Custom Sort Order
5. Change Page Orientation
6. Create Custom headers and footers
7. Save the spreadsheet

Sort Data

Work with Sample Data

Open the sample Excel list, Sales.XLS
When prompted, SAVE to your Documents folder

Review the Data

	A	B	C	D	E	F
1	Month	Client	Category	Service	Class	Date
2	January	Rick Towner	Private	Training	Access	1/12/2004
3	January	Darlene Davis	Private	Training	Access	1/15/2004
4	January	Hometown Community College	Educational	Training	Word	1/15/2004
5	February	Hometown Community College	Educational	Training	PowerPoint	02/05/2004
6	February	Harmony Kitchen And Bath	Corporate	Training	Excel	02/07/2004
7	February	Database Consultants	Corporate	Training	Access	02/10/2004
8	February	Bay County	Government	Training	Outlook	02/12/2004

Sort the Data

Select the entire spreadsheet and Sort the data by Month
  Did the Months sort as expected or did they sort alphabetically?
  Try the Sort again: use the CUSTOM SORT and change the Order to Custom List

Modify the Page Layout

Format the following Page Layout Options:
  Make the orientation "Landscape"
  Create a Custom Header and type a sample company a name in the center
  Create a Custom Footer with the current date on the right

Save the spreadsheet and name it: Excel Intermediate Practice 2

Intermediate Microsoft® Excel: Practice 3

Objectives:
The Learner will be able to:
1. Enter data into a Spreadsheet
2. Use AutoFill with labels, data and formulas
3. Format Cell Borders and Contents
4. Calculate the total across the rows
5. Calculate the total for each column
6. Use Conditional Formatting

Create a Time Sheet

	A	B	C	D	E	F	G
1	Monday	Tuesday	Wednesday	Thursday	Friday	Saturday	Total
2	8	8	8	8	8	8	48
3							
4							
5							
6							
7	8	8	8	8	8	8	48
8						Overtime	8

Enter the Labels in the first row
In Cell A1 type: Monday
Use the AutoFill handle to add Tuesday through Saturday

Calculate the Total
In Cell G1 type: Total
In Cell G2 create the equation: =Sum(A2:F2)
Use the AutoFill handle to fill down that equation to G6

Calculate the Daily Total
Enter sample data in cell A2 through F2
Select Cell A7 and AutoSum the total
Use the AutoFill handle to add this equation to Cells B7 through G7

Format the cells
Make the Labels Bold
Align all of the text Centered, in the middle of the cells

Calculate the overtime in Cell G8
The equation in cell G8 would be: =G7-40
Use Conditional Formatting on Cell G8

Save the spreadsheet and name it: Excel Intermediate Practice 3

Intermediate Microsoft® Excel: Practice 4

Objectives:
The Learner will be able to:
1. Enter data into an Excel Spreadsheet at least 75% of the time
2. Use Data Validation to create a DropDown Control at least 75% of the time
3. Rename a sheet in an Excel workbook at least 75% of the time
4. Apply Conditional Formatting at least 75% of the time
5. Use the Fill Down command
6. Enter data using Drop Down Controls at least 75% of the time

Create DropDown Controls

Create a list of employees
Type in the following information in Column A
- Bill Smith
- Kaylee Wild
- Helen Pulaski
- Corey Haas
- Angelique Riol

Select the data and name the range. In the Name Box type: Employees
Name the sheet: Employees

Create a list of locations on another spreadsheet in the same workbook
Enter the following Locations in Column A and sort them A-Z
- Ann Arbor
- Brighton
- Lansing
- Flint
- Detroit
- Grand Rapids
- Pontiac

Select the names and name the range. In the Name Box type: Location
Name the sheet: Location

Create a schedule on another spreadsheet in the same workbook
Rename Sheet3: Schedule
In Cell A1 Type: Employee Name
In Cell A2, use Data Validation to create a Drop Down control using the "Employees" source.
   Include an Input Message that says: "Select an employee from the list"
   Fill down Five rows
In Cell B1 Type: Location
In Cell B2, use Data Validation to create a Drop Down control using "Location" as the source
   Include and Input Message that says: "Select a Location from the list."
   Fill Down five rows

Apply Conditional Formatting
Fill in 3 rows of Employees with locations.
If the Location is Pontiac, format the text to be GREEN. Does Pontiac show up Green?

Save the spreadsheet and name it: Excel Intermediate Practice 4

;-)

Name: _____
Instructor: _____
Test Score: _____

Microsoft Skills Test

Intermediate Excel

☐ 1. Microsoft Excel Intermediate: Action Step 1
Open a new blank spreadsheet.
In cell A1 and type the word "April." Tab across to B1 and type "May." Use the Autofill to enter the labels from June through October

☐ 2. Microsoft Excel Intermediate: Action Step 2
Add the following values to your spreadsheet:
A2=3.4
B2=2.7
C2=1.8
D2=.75
E2=.5
F2=6.4
G2=3.4

☐ 3. Microsoft Excel Intermediate: Action Step 3
Select the range from A1 through G2. Name that range Rainfall.

☐ 4. Microsoft Excel Intermediate: Action Step 4
Create a graph using the Rainfall range. Select the Cylinder Chart type and remove the legend. Place the chart on the same spreadsheet as your data.

☐ 5. Microsoft Excel Intermediate: Action Step 5
Select the columns and double click to change the fill to red. Select the column for September and change the fill to yellow.

☐ 6. Microsoft Excel Intermediate: Action Step 6
Select the wall and change the fill to the Waterdrop texture.

☐ 7. Microsoft Excel Intermediate: Action Step 7
Double click the text on the axis and remove the Auto Scale. Format the type to be 9 pt.

☐ 8. Microsoft Excel Intermediate: Action Step 8
Rename the spreadsheet to be Rainfall Graph

Microsoft Assessment Test
Intermediate Excel

☐ 9. Microsoft Excel Intermediate: Action Step 9
Select sheet two and add the following labels:
A1=Expenses
A2=Rent
A3=Phone
A4=Electricity
Rename sheet two to be Expenses

☐ 10. Microsoft Excel Intermediate: Action Step 10
Select sheet two and add the following data:
B1=Amount
B2=1,500
B3=545
B4=130
Format the column for currency.

☐ 11. Microsoft Excel Intermediate: Action Step 11
Create a 3D pie chart from the Expense data.

☐ 12. Microsoft Excel Intermediate: Action Step 12
Change the Chart Options to show the Category name and the Percentage.

☐ 13. Microsoft Excel Intermediate: Action Step 13
Save the file as Your Name Excel Intermediate Chart Samples.
Please submit the spreadsheet to your instructor.

Name: _____

Instructor: _____

Test Score: _____

Microsoft Skills Test

Intermediate Excel

1. Microsoft Excel: Using Lists
A list has the column headers Last Name, First Name, Address, City, State, and Zip. In order to sort by City you would:

   - ☐ A. Select the City Column and go to Data ->Sort
   - ☐ B. Select the entire spreadsheet and go to Data ->Sort
   - ☐ C. Hide all but the City column then go to Format -> Data

2. Microsoft Excel: Using Lists
Check all that are true.

   - ☐ A. If the first row in your spreadsheet has labels, this is the Header Row. That's what Excel uses for the drop down boxes.
   - ☐ B. If you do not have a header, you will not be able to sort

3. Microsoft Excel: Form Design
There are several steps involved with creating a drop down list. Which of the following statements are true?

   - ☐ A. The Forms tool bar is turned on with View -> Tool bars ->Forms
   - ☐ B. Create a reference sheet with the information that will be displayed in the drop down list.
   - ☐ C. Select the Combo Box tool from the Forms tool bar
   - ☐ D. Drag and drop the size of the Combo Box
   - ☐ E. Right click the control and select Properties
   - ☐ F. Select the Input Range

4. Microsoft Excel: Form Design
Conditional formatting can be used to change the value in a cell to big, bold and red. Which of the following statements are true?

   - ☐ A. Conditional Formatting is added by going to Format -> Conditional Formatting
   - ☐ B. You can only set one Conditional Format at a time to a cell

5. Microsoft Excel: AutoFill
The data in Column A goes down to A20. If you double click the AutoFill on cell B2 the values will fill down to B20 automatically.

   - ☐ A. True
   - ☐ B. False

6. Microsoft Excel: AutoFill
Cell D2=100. Cell D3=105. Select D2 and D3 and double click the AutoFill handle. Excel will automatically add 5 to each new cell in the series.

   - ☐ A. True
   - ☐ B. False

Microsoft Assessment Test
Intermediate Excel

7. Microsoft Excel: AutoFill
Cell A2=Monday. Cell A3=Tuesday. How would you make Excel automatically fill in the weekdays?

- ☐ A. Select A2 and A3 and double click the AutoFill handle.
- ☐ B. Select A2 and A3 and drag down the AutoFill with the left mouse.
- ☐ C. Select A2 and A3 and drag down the AutoFill with the right mouse. Excel offers the option to fill in the days, or just the weekdays. .

8. Microsoft Excel: Formatting
Which of the following statements are true?

- ☐ A. To make the labels in Row 1 bold, select Row 1 and go to Format ->Font
- ☐ B. Double click on a worksheet tab to rename it
- ☐ C. To format a column for currency, go to Format ->Cells -> Number

9. Microsoft Excel: Formula
In this equation =Eggs!E6+Legs!E6+Pigs In A Basket!E6, Eggs, Legs, and Pigs in a Basket are the names of three spreadsheets in the same workbook.

- ☐ A. True
- ☐ B. False

10. Microsoft Excel: Formula
In this equation =Eggs!E6+Legs!E6+Pigs In A Basket!E6, Eggs!E6 refers to cell E6 on the Eggs spreadsheet.

- ☐ A. True
- ☐ B. False

11. Microsoft Excel: Formula
An Absolute reference, such as $A$2, means that the formula will point to cell A2 in the first equation, but B2 if you fill that equation to the next row down

- ☐ A. True
- ☐ B. False

12. Microsoft Excel: Formula
You can use Goal Seeking to set the value of one part of a spreadsheet and Excel will recalculate the other variables automatically

- ☐ A. True
- ☐ B. False

13. Microsoft Excel: Formula
What If Scenarios allow you to save different variables for your calculations

- ☐ A. True
- ☐ B. False

Microsoft Assessment Test

Intermediate Excel

14. Microsoft Excel: Formula
    To create a summary report of the differences between scenarios, go to Tools -> Scenarios -> Summary and select Report or Pivot Table.

    ☐ A. True
    ☐ B. False

15. Microsoft Excel: Formula
    $B3 is an example of a mixed cell reference, not relative or absolute.

    ☐ A. True
    ☐ B. False

Office Certification Excel Beginning Excel Intermediate Excel Advanced

Index Intermediate Microsoft Excel 2007 Exam 77-602 Guide

Autofit row height and column width, 5
Conditional formatting 55
 Rules Manager, 60
 Color scales, 58
 Data bars, 57
 Highlight, 44
 Icon sets, 59
 More than one rule, 56
 Top and bottom rules, 56
Copy a series, 73
Copy worksheets, 76
Data Validation: Data integrity, 49
Drop-down lists, 39
Fill a series without formatting, 29
Format all cells in a row or column, 72
Format rows and columns, 19
Format text: font, alignment, 48
Format the date, 69
Format: number formats, 72
Format: Cell borders, 32
Format: Create custom cell formats, 70
Format: Merge and split cells, 48
Formula: absolute references, 86
Formula: SUM, 30
Formula: Troubleshoot a formula, 79
Formula: Using data from other worksheet, 80

Insert a column or row above, below, to the left, or right, 31
Insert and delete cells, rows, and columns, 79
Insert and delete worksheets, 79
Insert multiple rows or columns simultaneously, 83
Pictures: Insert pictures, 46
Pictures: Modify pictures, 47
Print: Define the print area, 23
Print: Define the print area, 23
Print: Move a page break, 22
Print: Orientation, 21
Print: Scale to fit a printed page, 24
Print: Set print options, 23
Remove duplicate rows from spreadsheet, 41
Rename worksheets, 33
Rename worksheets, 76
Reposition worksheets, 76
Restrict data using data validation, 35
Restrict the type of data that can be entered, 38
Restrict the values entered, 39
Restrict to: Less than x, 50
Restrict to: Specified length, 50

Save to the Excel 97-2003 format, 63
Sort and filter data, 6
Table: Add and remove header rows, 12
Table: Add rows, 19
Table: Quick Styles, 15
Table: Band the rows or columns,16
Table: Change rows to columns, 16
Table: Change the total row function, 18
Table: Format the columns, 16
Table: Format data as a table, 8
Table: Insert and delete rows and columns, 19
Use Compatibility Checker, 63
View: normal, page layout, and page break preview, 20

The **Intermediate** Guide to **Microsoft Excel 2007**

Add Me To The List
Mail Merge with Excel

<u>Working Overtime</u>
Create a time sheet

<u>Legs, Eggs, and Pigs in a Basket</u>
Calculate Revenue

1. Guided Discovery
Step by step demonstrations of practical tasks.

2. Practice
Sample Spreadsheets

3. Assessments
Sample Skill Test
Sample Multiple Choice Test

Objectives
Exploring Strategies
Microsoft® Excel is a powerful tool for data analysis. This Guide will show how to use the data tools to filter, total and compare information.

Visualize: The average person will only spend three seconds reading your spreadsheet. Learn how to present that information quickly and effectively with charts and PivotTables.

What you will learn:
- Create and sort lists
- Use cell references and equations
- Format spreadsheets

Prerequisite knowledge:
Intermediate computer skills: knowledge of spreadsheet and table functions.

Length: The Intermediate Guide to Excel has 3 lesson. Each lesson can be done in 60 minutes or less.

Comma Project, LLC.
9090 Chilson Road
Brighton, MI 48116

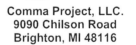
ISBN: 978-0-9818778-5-3

9 780981 877853